T0331067

Interactive Sound and Music

Interactive Sound and Music: Beyond Pressing Play provides an accessible exploration into the aesthetics of interactive audio, using examples from video games, experimental music and participatory theatre and sound installations. Offering a practitioner's perspective, the book places interactive sound and music within a broader aesthetic context relating to key texts and discussion within musicology and wider art practices. Each chapter takes the reader through a key debate surrounding interactive sound and music, such as:

- Is it actually interactive and does it actually matter?
- How do audience expectations change in an interactive space?
- How do you compose for multiple possibilities?
- Is interactive sound and music ever finished?
- Where now for interactive sound and music?

Supported by a series of questions at the end of each chapter that can be used as a focus for seminar or reading group activities, this is an ideal textbook for students on audio engineering, music technology and game audio courses, as well as an essential guide for anyone interested in interactive sound and music.

Lucy Ann Harrison is a composer, sound designer and academic based at the University of Westminster, UK. Works include a library themed puzzle game, an interactive blanket fort complete with a musical hopscotch, motion sensitive sound created using gaming controllers and critically acclaimed music for immersive theatre.

Interactive Sound and Music

Beyond Pressing Play

Lucy Ann Harrison

Routledge
Taylor & Francis Group

LONDON AND NEW YORK

Designed cover image: Sorapop via Getty Images

First published 2025
by Routledge
4 Park Square, Milton Park, Abingdon, Oxon OX14 4RN

and by Routledge
605 Third Avenue, New York, NY 10158

Routledge is an imprint of the Taylor & Francis Group, an informa business

British Library Cataloguing-in-Publication Data
A catalogue record for this book is available from the British Library

Library of Congress Cataloging-in-Publication Data
A catalog record has been requested for this book

ISBN: 978-1-032-38240-1 (hbk)
ISBN: 978-1-032-38239-5 (pbk)
ISBN: 978-1-003-34414-8 (ebk)

DOI: 10.4324/9781003344148

Typeset in Sabon
by Taylor & Francis Books

Contents

Illustrations

Figures

Tables

Boxes

Preface

Interactive sound and music represents an interesting challenge for the composer. The work is ephemeral and changes continuously. It requires the composer to think in a non-linear fashion and consider how audiences, will respond and behave. The composer has to create the work based on the potential possibilities that could be produced rather than from a fixed, knowable output.

I first began working with interactive sound and music as a postgraduate when I learnt the software Max MSP. Prior to this point, I had been struggling with the restrictions of linear composition and how the behaviour expected in concert performances impacted my desired audiences' behaviours. I found the concert imposed a formality on my work which led to my intended audience feeling like they were judged if they clapped in the wrong place or behaved in the wrong way. There was a disconnect in the feeling that I wished to create for my audience and the format that I was working in. Moving to interactive sound and music opened all aspects of my work up. It allowed me to better realise my compositional aims, moving to more flexible work and open structures that allowed my audience the chance to play and explore.

With this book I have a similar aim, represented through the discussion of the aesthetics of interactive sound and music. The book shows the infinite possibilities of interactive sound and how it can lead to a collaboration between the audience and composer who are united through the potentials of the work. This book aims to address key questions in interactive sound and music in a way that is accessible and demystifies the composer's approach.

Acknowledgements

Thank you to my family, friends and colleagues who supported me during this writing process. Particularly those who listened to me process thoughts and approaches through dinners, coffees, phone calls and group texts.

Introduction

About this book

This book offers an introductory and accessible guide to the aesthetics of interactive sound and music, considering the technical, social and artistic requirements of the work and how these influence the composer's decisions and the sound being created. Often these decisions are based on embodied knowledge developed by the composer through their practice which have become instinct; this book will use my experience as a composer and sound designer to attempt to verbalise those practices.

Within the field of research into interactive sound and music, aesthetic debates are usually separated by media type, for example there are many excellent books that focus on video games only as a field, some of which will be referenced later in the text. This book will provide an overview which compares the approaches across game audio, interactive sound installations, experimental sound and music and public art allowing the reader to consider the similarities and contrasts across each form and the impact this has on the aesthetics of the work.

This book considers the role of both the composer and the audience, looking at how interactive approaches impact the hierarchies of the composer as they work towards a long-distance collaboration with the audience.

Each chapter is framed around a question related to aesthetics and links to wider debates not only in music but in other disciplines including immersive theatre and exhibition design reflecting the cross-disciplinary nature of interactive sound and music. While the chapters do not offer definitive answers, they aim to open up debates and further questions about the aesthetics of interactive sound and music.

At the end of each chapter there are a series of questions to support reading group study. These are intended to provide additional areas of enquiry or study related to interactive sound and music from a composition, audience experience and analytical perspective.

DOI: 10.4324/9781003344148-1

Outline of chapters

Introduction: What do histories tell us about interactive sound and music?

This introductory chapter provides an introduction to the aesthetic values of interactive sound and music through investigating the written histories of game audio, experimental interactive audio and public art and interactive sound installations. Written histories, interviews and company approaches are analysed, considering strands that are used within the histories in order to determine how the histories and approaches are becoming solidified and developing into a set of aesthetic values for each approach. This will be used to consider the similarities and differences of each field's approaches to interactive sound and music, building a basis for the discussions underpinning the debates and considerations for future chapters.

Chapter 1: Is it actually interactive and does it actually matter?

When working within an interactive audio setting, the composer makes a choice about how much control they retain over their work and how much they make interactive. This can impact the audience's engagement with an interactive work and the overall aesthetic of the finished output. If you keep too much control over the work the final product will sound exactly as the composer intended but may seem superficial or unsatisfying to an audience who has been sold an interactive experience. Give up too much control and the audience will have a truly interactive experience which may seem messy or lacking a clear vision. This chapter investigates where the balance lies for a composer when creating an interactive piece and the impact on audience experience. It provides a scale of interactivity for sound and music, considering the ethical and social obligations that a composer has towards the audience within an interactive work.

Chapter 2: How do audience expectations change in an interactive space?

When audiences approach an interactive space they have a different set of expectations based on their prior understandings and experiences of interaction. This chapter considers how these approaches impact an audience's understanding of the work and how these change based on confidence with interaction and group dynamics in a space, building on key theories such as Chion's modes of listening,[1] which outlines how an audience listens differently based on the context, Small's concepts of Musicking[2] and user design techniques such as Quantic Foundry's gaming motivations.[3] Through this lens the chapter will explore the importance of intentionality in a medium where the audience will subconsciously or consciously read into all decisions that the composer or sound designer makes.

Chapter 3: How do you compose for multiple possibilities?

This chapter addresses how sound and composition processes change when working in an interactive format. The chapter is divided into two sections. The first section addresses the role of the sound designer using the audio roles of immerse and inform to provide a practical framework to consider audio as an interface and a tool for narrative immersion. Practical solutions are offered using the author's own professional experience and placed within wider aesthetic frameworks. The second section looks at musical composition processes, moving from structured composition processes used in game and narrative work through to freer structures that are more suited to gallery or public art environments. Composition processes are discussed using practical examples from the author's own work as a composer, relating these to other composition practices in harmony and counterpoint and 20th century composition.

Chapter 4: More than novelty: How can you make interaction a meaningful part of the work?

When creating interactive sound and music the composer is responsible for ensuring that the work is a real collaboration between them and the audience, while considering the interactive need of the work. For interaction to be meaningful it needs to be fully embedded within the compositional process and the outcomes of their interaction must be visible to the audience. This chapter builds on previous discussions on the authenticity of interaction, audience expectations and practical approaches to interaction using examples from games, installations and the composer's own experiences to develop approaches to ensure that interaction within the work is meaningful.

Chapter 5: Is interactive sound and music ever finished?

When working in a medium where the result changes in every performance or interaction, the composer or sound designer working with interactive audio is often working within a frame where they will never see the fully finished output. They can test different iterations but once the work is out in the world it takes on a life beyond the control of the composer. This chapter investigates whether we can ever consider a piece of work complete if it relies on a member of the audience to interact with the work. It builds on approaches to completeness from music and art considering the impact of musical structure, audience roles and composer intentions on completeness in interactive sound and music.

Chapter 6: Can we create an artefact of a piece of interactive audio?

This chapter addresses debates around archiving and creating an artefact of interactive sound and music. Reasons for creating an artefact can include recreating the work in the future, providing a record of the work for analysis and study, advertising a work and preserving and archiving work that is considered culturally significant. As the field is now at the point where interactive sound and music is being added to archives and the canon is being established, decisions being made today will impact our archives and stored material for future analysis and understanding of the work and study. The chapter will look at the concept of the work in the context of interactive sound and music, drawing on established aesthetic debates to provide workable approaches to create and artefact for a range of interactive sound and music.

Chapter 7: Where now for interactive sound and music?

To conclude, this chapter considers current trends in interactive experiences and technology, providing a speculatory view of how changes in interactive sound and music will impact the roles of the composer and audience moving forward. It will provide suggestions about how approaches may become embedded across different fields and will consider the framings of games, participatory art and experimental music and whether these worlds will remain separated into the future.

By examining these aspects from an aesthetic standpoint and contributing a practice-based perspective, I aim to provide an insight into how the work is built with the considerations that sit behind a composer's intentions, reflecting how the aesthetics of interactive sound and music function.

What do histories tell us about interactive sound and music?

In order to begin discussing the histories of interactive sound and music, it's important to understand the historical contexts and the impact that these have on aesthetic values.

Within this book interactive sound and music sits in three different areas:

- Sound and music for games, where the interaction, broadly speaking, has a functional or applied need (although the sound and music can also have an artistic function, and in many of the best games does)
- Experimental sound and music, where the interaction forms part of the artistic intention or expression of the work
- Immersive events or public arts which hold both a functional and artistic role

In general, these three areas of interactive sound and music are discussed separately. This could simply be due to how the areas were developed and the different audience types in each field. However, by considering the histories and developments in each area of interactive sound and music separately we run the risk of missing similarities between the areas, aesthetics considerations and technologies used e.g. Max for Live is now embedded in Ableton Live and is a commonly used tool in live performance practices in commercial music, however it can sometimes be discussed without reference to its full counterpart Max, which has roots in experimental sound and music. Further to this, discussing the areas separately without drawing links between the different practices creates potential narratives of 'high' and 'low' culture such as we already see in classical and popular music discussion, where one field is seen as 'complex' and another is seen as 'accessible'. This does a disservice to all areas involved in these narratives.

The way that histories of interactive sound and music are told is pivotal to how the different areas are perceived and the aesthetic considerations in each area. As with the telling of any history, the areas that we choose to emphasise or omit point to the value system and the culture around the history and this becomes more solidified in each retelling, leading to the canon which emphasises the 'greats'. We are at an interesting point in interactive sound and music, where the practices and approaches are still relatively new. This means that for the three areas of games, interactive experimental sound and music and immersive and public art, the 'canon' has not yet been set but the histories are beginning to be told; we are able to see how the legends surrounding the areas are being developed.

While many other books have provided a comprehensive oversight of the history of interactive audio technology, this chapter will be considering how the different stories are told within each field and how these point to aesthetic values and areas of overlap between the different fields of interactive sound and music. This will include looking at interviews, published academic histories and company mission statements and values. Through this approach links will be drawn between different approaches, considering what this means for the sound and music being created and for the audience or player experience.

Games

The history of music and sound for games can be started at various points in history.

In the pivotal work 'Game Sound: An Introduction to the History, Theory, and Practice of Video Game Music and Sound Design',[4] KC Collins begins her history of game sound with mechanical arcades and casinos with bagatelle, pinball machines and one-armed bandits, looking at how the sounds made by bells and ball bearings impact a player's understanding of gameplay and encourages them to continue playing. This provided an early semiotic association with a bell *ding* as a positive, matched in games such as Mario with the coin collection sounds and later mimicked by the default iPhone text tone.[5] The use of a ball bearing hitting the bell is a very practical solution to providing positive feedback in a game.

This begins the consideration of games as a practical, solution focused area of interactive sound and music which we see underpinning all histories and 'legends' of sound in games. The practical solution found in pinball machines, looking to provide instant feedback on gameplay, shows that game audio as an interface is a central concern, and the simpler solutions are favoured as they will provide simple solutions and easier link to player understanding.

Happy accidents in sound

The history of games then develops into a story about experimentation and technological limitations with a practical focus at the centre. Early computer games had limited sound capabilities, the memory in early games was minimal and sound files can take up large amounts of space. However, feedback was needed to improve gameplay. A commonly used example of this in video game history is the sound for the 1972 Atari game *Pong*, where the circuit was hacked to create the iconic beep sounds.

Computer engineer Allan Alcorn stated in an interview with NPR:

> As a solution I just poked around in the circuit itself, in the vertical shape (ph) generator for appropriate sounds and piped them out... remember; this was 1972. There was no internet. There was nothing. And so I just poked around. And it took me about two hours.[6]

This highlights how simple, practical solutions are valued within game audio, but also suggests that this simplicity and 'not overthinking the solution' is prized in the field. The phrases such as 'just poked around' and the emphasis on how quickly the solution was found point to the practicality of working in an industry with rapid deadlines but also suggest something that puts this field at odds with more experimental sound and music creators that favour process, even though experimentation sits at the heart of both approaches (it's also noteworthy that the creator of one of the most iconic early games sounds was not an audio specialist, but an engineer).

Supposed happy accidents, such as those ascribed to *Pong*, are central to the narratives in games audio development. *Space Invaders*, another Atari game from 1978 with iconic sound, increases the tension in the game due to the increasing tempo of the music as the lines of spaceships drop. By this stage, the process of hacking the circuit to develop sound had become more sophisticated to the point that designers could produce different timbres and sound types. *Space Invaders* has sounds for shooting, enemy missiles, the boss alien and (some) musical underscore. The increasing tempo of the music causes players to act with more haste and potentially make more mistakes as they work through the game – it ups the stakes, making the game more fun. However, this increase in tempo was caused by a bug in the game which ended up enhancing game play, as Bill Adams, director of game manufacturing for Midway Games once stated:

> The speeding up of the space invaders was just a function of the way the machine worked ... The hardware had a limitation—it could only move 24 objects efficiently. Once some of the invaders got shot, the hardware did not have as many objects to move, and the remaining invaders sped up.[7]

Again, the narrative around Space Invaders is linked to practicalities of technologies and gameplay experience supporting the player understanding of game mechanics. It doesn't matter if something is a happy accident if it supports and improves the game play.

These early uses of sound in games shows the priorities for sound in games as:

- Sound's role for understanding games (audio as an interface) demonstrated through practical and easy to follow solutions built on easy-to-understand semiotic associations
- Sound as a tool to influence audience behaviour – in the case of the pinball example the positive association of the bell leads to repeat gameplay and for *Space Invaders* the increasing tempo leads to further mistakes in gameplay and greater excitement

Sound developing into a serious concern in games

One of the most notable early composers in game music is Hirokazu 'Hip' Tanaka, an early pioneer of chiptune as a genre. Tanaka worked with Nintendo from 1980 on titles such as *Metroid, Donkey Kong* (when Mario was still called Jumpman) and *Duck Hunt*. Tanaka described the shift for game music being considered an art form in its own right, stating in an interview:

> The sound for games used to be regarded just as an effect, but I think it was around the time *Metroid* was in development when the sound started gaining more respect and began to be properly called game music. Even the media had put spotlights on it, and we began to see many articles on game music.
>
> Then, sound designers in many studios started to compete with each other by creating upbeat melodies for game music. The pop-like, lilting tunes were everywhere.
>
> The industry was delighted, but on the contrary, I wasn't happy with the trend, because those melodies weren't necessarily matched with the tastes and atmospheres that the games originally had.
>
> The sound design for *Metroid* was, therefore, intended to be the antithesis for that trend. I had a concept that the music for *Metroid* should be created not as game music, but as music the players feel as if they were encountering a living creature. I wanted to create the sound without any distinctions between music and sound effects. The image I had was, "Anything that comes out from the game is the sound that game makes."[8]

This quote from Tanaka shows the beginning of a shift for game audio which looks at how interactive sound and music in games can be used practically to support game play as well as artistically as scoring for the narrative; this further links to current game practices where we have moved to more orchestral scores and cinematic scoring approaches.

Throughout the 80s game audio developed in line with technologies that mirrored the development of synthesisers, chips had greater capacity. The move to 16-bit chips in the 80s allowed for wider incorporation of FM synthesis into games. This supported specific synthesised instruments to be represented in the music and more realistic and varied sounds within games. Alongside the increased complexity of music in games were chips that allowed limited speech within a game. Collins discusses this within an arcade context where 'the machines would literally call out to players, begging to be played'.[9] Again, this emphasises a player focus alongside the technological development encouraging the players to engage with the game.

While these developments had been discussed in an arcade games context the move to home consoles also emulated the 16-bit developments, through consoles such as the Sega Mega Drive (Sega Genesis in North America). Collins, again, highlights the practicalities of the technology with programmed sounds being shared between games.[10] These stories point to ingenuity but also the practical needs of working with developments and technological limitations in games.

Finally, with the development of more powerful home consoles such as the Playstation, which had a sound card built in, composers moved to orchestral scores as seen in contemporary AAA games. This led to increased flexibility in games composition, with composers from film now crossing over into games and bringing over cinematic practices, most notably Hans Zimmer in *Call of Duty: Modern Warfare 2*.[11] Zimmer considered the interactivity within the games to be a benefit of the form, saying in an interview with USA Today: 'as a musician, we play all our lives, so the idea of playing something and being involved in something is actually quite powerful to a musician. The participation is the thing'.[12]

This aligns with the development of more cinematic games, as pioneered by developers such as Hideo Kojima.

Essentially, through the story of interactive audio in games we get a history of technological limitations and ingenuity which provide the principle for how game audio works. The development of technology is emphasised throughout game audio histories and is used to show how technology is linked to changes in game sound practices.

In the early stages engineers were favoured above audio specialists and the practicalities of the sound were focused on providing more clarity through gameplay. Through the technological limitations best practice was set up to prioritise audio as UI/UX. It is expected that players will read into the audio being created and make choices based on the sound.

Within this narrative the player or audience sits at the centre of the work. All sound is developed to support their play or experience of the game. It instructs them how to play, gives real time feedback and improves the immersion of the player experience. Once this approach had been developed we see the rise of the composer, led by the work of pioneers like Tanaka. Through this, game sound and music moved towards a more cinematic approach with music being developed to support the narrative and immersion in the game.

The role of training

Aesthetically, through the narrative of histories the priority is always the player and their experience, whether this is the through the UI/UX use of audio or through a more narrative/cinematic focus.

Away from the player priority the other key part of the stories told around game audio is the happy accidents, the experimentation and DIY approach of it all. Training is downplayed in favour to how quick the solutions were and how tight the turnaround for development, as in the solution found for the sound in *Pong*. As Collins states:

> The social constraints of specialized knowledge required to program the software—and to develop audio and write music—also influenced the developing aesthetic, with many composers lacking formal musical training, which is not to say lacking in talent. Even when musicians were involved, they were generally separated completely from the integration of the music.[13]

What this suggests is that, even though games are a multi-billion-pound mainstream industry, there is an attachment to the narrative of the outsider and the counterculture, the same as with Apple still using the narrative of Jobs and Wozniak working out of a garage. Through this they build a sense of community, of pioneering outsiders. For sound and music, this suggests that solutions can be found not through using traditional musical training but through looking at the technology first and working out a solution. It suggests that the technology is the part that needs to be mastered and the sound is instinct.

Key themes in game audio histories

Based on the histories as outlined above there are the following aesthetic themes to be considered when working on sound and music for games.

Game audio is practical and audience focused. From the beginning we see a focus on how the audio can help a player understand gameplay, particularly through use of semiotic associations in game audio and the role of audio as an interface.[14]

Practical approaches are favoured to create innovation. As the industry works often with technological limitations around storage and sound capabilities, these practical approaches are emphasised throughout the histories. Those working with game audio are expected to be solution focused and work creatively with technological limitations.

Happy accidents are prioritised within the histories showing that creators feel there is a benefit to sound being seen as something instinctual which is understood by those working in games. Genius is seen to be innate and linked to compositional and audio instincts held by some individuals in the field.

Experimental interactive music

Histories of experimental, interactive music usually take The Stanford Center for Computer Research in Music and Acoustics (CCRMA) as a starting point and expand out to similar models for development, most notably IRCAM (Institut de recherche et coordination acoustique/musique) in Paris.

CCRMA was formed in 1975, with its origins in the 1960s, through the work of composer John Chowning and his colleagues who borrowed and repurposed equipment from Stanford's Artificial Intelligence Laboratory. The centre has provided a template for other experimental computer music organisations such as IRCAM and set the field's standards for experimentation, collaboration and aesthetics considerations.

While the history of game audio focused on the practical requirements of the game and the player experience, the narrative of experimental interactive music from organisations such as CCRMA and IRCAM focuses on their role as pioneers, technological innovation, collaboration and musical development.

Music and technology

In CCRMA's account of their history they begin in 1964 where John Chowning was studying with Professor Leland Smith, and Nelson also emphasises in their history of CCRMA Chowning's studies with Nadia Boulanger.[15] He developed collaborative partnerships with Max Mathews of Bell Telephone Laboratories and computer scientists David Poole.[16] This sets up two priorities when discussing experimental interactive sound and music. Firstly, it links Chowning to experimental composition practices through Leland Smith who, in turn, studied with Darius Milhaud and Olivier Messiaen. Similarly, histories of IRCAM, founded in 1977, centre on Pierre Boulez who was invited to return to France by President Pompidou in order to found and direct IRCAM.[17] These compositional approaches are key to the focus of these histories as it links CCRMA to both the experimental avant-garde and traditional musical training being applied to experimental computer music. Music created from these centres is expected to reference and build on past compositional practices. As Boulez states in his lecture on invention and research at the opening of IRCAM, 'creativity does not come from nowhere but is nurtured by contact with music of the past'.[18] Musical training is raised as being important for the development of new work. While some elements of the avant-garde have aimed to distance themselves entirely from references to the past, this isn't possible. As is a common truism across the arts, one must know the rules in order to break the rules.

Collaboration between musicians and technologists is central to the histories of experimental computer music and is reinforced from the beginning of the CCRMA, where technology was borrowed from the Stanford Artificial Intelligence Laboratory (SAIL). Max Mathews, who worked in the Bell Laboratories acoustic research department, is a common link between different computer music centres. Mathews wrote an early computer music article read by Chowning about computer music which instilled the idea of computers as a general-purpose device.[19] Chowning then visited Mathews, who provided Chowning with guidance and programmes to support his work. Mathews was also IRCAM's first scientific director in 1975, providing links between IRCAM and Bell Labs. The collaboration with Bell Labs is central to the work of CCRMA and also provides an important grounding for the field-aligning experimental practice both to industry and to the Avant Garde. Bell Labs' Experiments in Arts and Technology collective, formed in 1966, was a key part of the Bell Lab's role in experimental electronic music and facilitated collaboration between artists and engineers including John Cage. Julie Martin's history in *IEEE Potentials* raises the importance for Bell Labs to ensure that their engineers understand the social implications for their system through collaborating with artists.[20]

Physical environment

Nelson raises the importance of the physical environment to the collaboration process:

> The researchers' relative isolation in the foothills outside the main campus ensured that they interacted frequently; although their background and interests were diverse, they were the only people around, and a certain level of interaction was almost unavoidable.[21]

Georgina Born also focuses on the importance of the building in her writing on IRCAM, highlighting instead the isolation of the working processes which are punctuated through 'informal, spontaneous oral consultation, as well as sustained collaboration'.[22]

However, practices at IRCAM have evolved since Born's observations in 1995 and collaborative practices within the building are central to the principles outlined on IRCAM's website:

- **Research:** 'The fundamental principle of IRCAM is to encourage productive interaction among scientific research, technological developments, and contemporary music production. Since its establishment in 1977, this initiative has provided the foundation for the institute's activities.'[23] This directly links to Boulez's requirement of scientists and composers working collaboratively
- **Creation:** 'IRCAM is an internationally recognized research center dedicated to creating new technologies for music. The institute offers a unique experimental environment where composers strive to enlarge their musical experience through the concepts expressed in new technologies'[24]

Collaborations with technology were mutually beneficial to both the technologists and the creatives. Composers and artists were able to build out their creative vision while technologists were able build and test commercially viable technology. In *The Sound of Innovation* Nelson and Mody refer to this as 'radical interdisciplinarity'[25] built on the concept that neither art nor technology is raised within the hierarchy of the work, they are considered equal in the collaboration. As Beck and Bishop state, 'interdisciplinary networks provided a context within which individual capacities might be recognized while rejecting the idea that any one individual may be enough'.[26]

This is reinforced by Pierre Boulez in multiple places in his writing about establishing IRCAM: 'technological research has often been the work of scientists who, while interested in music, stand outside the usual circles of musical education and culture'[27] while 'the most adventurous musical spirits dreamed of using technology to other ends. Through an intuition that was both certain about direction and uncertain its implications, they imagined technology could be useful in the search for new material'.[28]

It is important here that these two groups find a way to work together and speak a shared language.

> The inability of many scientists to understand exactly what musicians want from them, and their failure to identify possible fields of cooperation, leads them to send the absurdity of the situation. It's as if a sorcerer were forced to beg for the services of a plumber![29]

This highlights that a priority in this field is the composition process. Secondly, it establishes the importance of industry and cross discipline collaboration to CCRMA.

In *Technocrats of the Imagination*, Beck and Bishop state that:

> The postwar emphasis on interdisciplinary collaboration and creativity in the human and physical sciences chimed with the broad contours of the historical avant-garde's challenge to bourgeois art, namely through a rejection of individual genius and a stress on collective practice.[30]

However, this is contradicted by the histories that have been written on the different centres in music. The central role of Max Mathews and his influence on computer music practices is written about across all histories, with Georgina Born referring to him as the 'father of computer music'[31] whilst making Boulez the focus of her history. Nelson devotes a large amount of his history to the role of Chowning, including discussions on his tenure review case and dismissal.[32] Similarly, within *Technocrats of the Imagination*, even while stating that the focus was not on individual genius and the writing emphasises a rejection of genius throughout, there is also a strong focus on individual creators and their processes.

Born discusses this contradiction within the practicalities of IRCAM as an institution where, at the time of writing, while considered higher in status to the aims of the institution, composers were paid less than those in technical or administrative roles as, excepting established stars, 'IRCAM did not need them. Rather they needed IRCAM for prestige and career advancement'.[33]

This contradiction is significant in the discussion of experimental interactive music as a focus on a singular 'genius' within the history, something that is common within classical practices and in the development of the canon. However, the notion of the collective is important to how the histories of experimental sound and music are written.

> The collectivist dimension of art-and-technology collaborations is a vital part of their utopian promise since it represents an attempt by many artists and organizers to instantiate radical democratic practices, or at least to start to imagine what such practices might be, within the ostensibly hierarchical and corporate environment of the mid-twentieth-century university or business.[34]

Even if coverage and histories of the work centre a big 'name' or a 'singular genius', the field itself values collaborative practice that put the art and the technology on an equal footing. This suggests that there is no hierarchy between the music and the technology that facilitates interactive processes. These exist in a symbiotic relationship where the two fields and processes are reliant on each other to develop and enhance their practices.

Musical exploration and freedom

While discussions of music and technology have generally been placed on an even footing throughout the published histories, there is still an expectation that composers will be able to be exploratory in their approach.

The creators of Max MSP, a visual programming software commonly used to facilitate interactive sound and music, Puckette and Zicarelli, both discuss the 'blank' page element of Max and Pure Data as being central to the functions of the software, with Puckette stating that this blank sheet of paper is the antithesis of other software that start from a template that pushes one style of musical development:[35]

> ...we software writers must try not to project our own musical ideas through the software ... To succeed as computer music software writers, then, we need close exposure to high-caliber artists representing a wide variety of concerns. Only then can we can [sic] identify features that can solve a variety of different problems when in the hands of very different artists.[36]

And:

> The whole business of the software hegemonists is to impose context in the form of proprietary file formats, operating system features, and other such constructs. Their concern is not in letting people do their own thing, and their products will always impede—rather than encourage—progress.[37]

While Zicarelli states:

> One of the things that I have always liked about Max is that it has no musical concepts in it at all… I think that introducing concepts that explicitly had to do with musical structures into the fundamental core of the environment would be a disaster.[38]

This suggests that the creators expect a software like Max to not dictate the work of the composer, instead providing more flexibility for open, musical exploration.

Boulez also warned about processes and practices that adhered too much to historical practices in place of more free explorations: 'musical life thus assumes the character of a museum, with its almost obsessive concern with reconstructing conditions of the past as faithfully as possible'.[39] While the music is informed by the past, this can become constrictive and limiting, particularly in the choice of instruments and timbres. Creativity should not be hampered by current limitations of technology and realities to develop the imagined work.

Within experimental, interactive sound and music the composer is expected to work outside of the expected and existing structures. They should be striving for new ways of working and new sounds, supported by the technological developments. While the composers working in this field historically are linked to schools of training and compositional develop, with experimental composers having undertaken formal musical training, the expectation is that composers will use the blank pages of the tools and their collaborations with technologists to push musical forms away from traditional structures.

Musical and technological pioneers

A key theme across histories in experimental computer music is the role of institutions and creators as pioneers. Emphasis is placed on technological developments, patents and being the first to work in certain processes.

Within the CCRMA's history they particularly emphasise the pioneering nature of the research and their role as the pioneers in the field. They highlight the following 'firsts' for the sector:

- The first on-line computer composition and synthesis system
- The first programs for moving sound through a four-speaker space
- The first course in computer-generated music (1966)

- The discovery of the use of frequency modulation (1967)
- The first high quality digital recordings (1976)[40]

The same emphasis on firsts is used in Nokia Bell Labs' history of The Experiments in Arts and Technology collective which states that the group was responsible for 'the first large-scale collaboration between artists and research engineers',[41] while IRCAM's timeline outlines a range of software innovations including the 4X system, Max, Modalys, spatializer software and Audiosculpt.[42]

By emphasising firsts in the fields we show a priority for innovation and research within experimental interactive music which links to the collaborative work and roles of technologists. It is important for the field that CCRMA is seen as a centre for innovation, and that they establish the commerciality of the inventions that were created. These experimental organisations sit within practice-based research communities and are dependent on research and patent funding, while IRCAM's pioneering research demonstrates a commercial approach to their pioneering technology that sits alongside their avant-garde compositional leanings.

CCRMA also emphasise within their history their role advising other organisations, most notably being advisers to Boulez in the establishment of IRCAM.[43] Among this support included a meeting with notable experimental composers. This advisory role led to a number of Stanford graduates and former Stanford and Bell Labs personnel working in key roles at IRCAM, a decision linked to the presence of Max Mathews. While reliant on these colleagues for technological expertise, the presence of American personnel within IRCAM led to a number of conflicts within working processes and aesthetics. Born writes that visiting Americans found IRCAM's environment to be 'unprofessional and beaurocratic'[44] and reports on a number of conflicts between Boulez and Mathews, as well as conflicts between Boulez and other American colleagues. These conflicts included the types of technology being developed, Boulez's stating that a composer's work was 'lightweight' and dismissing Mathews' work on a fretted violin.[45] Through highlighting these particular conflicts between the American researchers and their colleagues, particularly Boulez, Born signals that while the two organisations came from the same source, there are different aesthetic philosophies underpinning the work and IRCAM was not aiming to directly replicate CCRMA, but to build on the approach with their own concerns, centralising Boulez in these discussions about direction and aesthetic centres, his work and experimental compositional processes.

The conflict between open source and commercial

Due to the innovative aims that have led the institutions to be labelled as pioneers, within the collaborative artistic and technological innovation at each institution a large amount of profitable intellectual property has been developed.

A key thread in all histories is the conflict between open-source ideals and commercial values.

While Chowning was working at Stanford, the university were beginning to develop their Office of Technology Licensing (OTL), which became the 'Stanford Model' of technology transfer, led by Niels J. Reimers. Interviewed for the Stanford Oral History Collection, Reimers stated the aims of the OTL, including better negotiation for researcher skills and more profitable use of patents.[46] Chowning chose to disclose technology for sound localisation to OTL with an aim to allow wider distribution of the product. Nelson described Chowning as aiming for a 'collaboration with a commercial firm',[47] extending the perception of the collaborative goals of CCRMA.

Chowning began work on FM synthesis while at CCRMA, undertaking the initial work in 1967. As with games histories, Chowning describes this as happy accident of experimentation: 'It wasn't an invention as much as it was a discovery. It was always there. And ever since, it's been full of surprises for everybody who has ever used FM'.[48]

CCRMA's patent for FM synthesis was granted in 1977[49] and OTL licensed Chowning's technology to Yamaha. Reimers described the negotiation and development process as taking about four years to complete, during which time the technology was copied by other companies, leading to multiple sub-licensing deals.[50]

The open-source, collaborative ideals of CCRMA provided some conflict with Chowning who shared information from the patent with colleagues at other institutions and discussed the technology openly in interviews, noting the conflict between the university's openness and Yamaha's corporate approach to not disclosing information: 'in the Stanford-Yamaha case, collaboration across "science" and "technology" led to misunderstandings and apparent missteps'.[51]

A similar challenge exists within the history of IRCAM, which has generated a range of viable commercial technologies. As these began to be developed in 1984, Born described a 'hostility and contempt towards all commercial developments and especially "low-tech" or small consumer technologies' that extended from Boulez's artistic ideologies that held no interest in commercial ventures.[52]

While this hostility is reported within the history of IRCAM, the organisation is responsible for the highly commercial software Max.[53] Originally developed by Miller Puckette in 1988, Max was further developed to include signal processing by incorporating developments by David Zicarelli. In an interview in the *Art + Music + Technology* podcast Puckette discussed the original development of the technology:

> I actually didn't set out to make a visual programming language at all. But in fact, what I wanted to make was a real time scheduler that could, that could be used in a musical situations.[54]

Again, similar to Chowning and game audio origins, this emphasises happy accidents in developing innovations.

From here we see a divide in the development of Max MSP between commercial and open source approaches. Miller Puckette left the company, the software continued to be developed by David Zicarelli as part of Cycling 74. This development through Cycling 74 had a very strong commercial approach. In a panel at the Dartmouth Symposium on the Future of Computer Music Software in 2002 Zacarelli outlined two aims for the development of Max:

> ... one is that, because the API is relatively simple and well-documented, it should be possible to incorporate other software into it as an object, such as Csound or even the SuperCollider language, and the other thing is that Max/MSP is eventually going to be a library that can be used for the purpose of customizing other software.[55]

We can see the culmination of these aims in the incorporation of video tools such as Jitter into the Max visual programming framework, and the embedding of Max for Live into Ableton Live.

Meanwhile, in 1996, Max's creator Miller Puckette launched an open-source visual programming framework, Pure Data, which works using a programming and processing framework which pushes a more community focus to the work, with open source development and instructional guides provided using a community wiki linking to Puckette's belief, outlined in his review of Max at 17, that 'communities are necessary for knowledge to grow' and that we should work to democratise computing, while acknowledging that selling software is also necessary.[56]

This conflict between open-source and commercial goals points towards a central challenge of collaborations between commercial and academic or arts organisations. The ideal of free and open sharing of knowledge needs to be balanced with the commercial challenges of industry and ensuring that developments are self-sustaining in a difficult funding environment.

Key themes in experimental sound and music

Based on the elements emphasised in experimental computer music, the following elements are core to the aesthetics of experimental interactive sound and music.

Avant-garde composition sensibilities can be seen throughout the histories, most prominently through the role of Boulez in establishing and directing IRCAM. This leads to an expectation of more experimental practices to be built on subverting classical traditions and aesthetics. Through the use of the technology, these compositions are expected to break away from traditional compositional practices, forms and processes.

Collaborations between technologists and composers are central to the development process of experimental computer music practices. Within this process the composition and technology are considered to be equal in importance; in the development of interactive work these components should be treated equally. Within the field there is a perceived value to taking technological risks.

Open-source is an ideal within the community but this conflicts with the wider distribution aims of software and practices and the ability to fund future developments and projects.

Public art and interactive installations

Public interactive art is seen within either a gallery, public space or participatory context. These contexts can overlap, for example gallery and public work can also be participatory. Participatory work could exist in a gallery. Broadly speaking, I am using this category to refer to interactive installations – sound works and sculptures that exist in public spaces and promote audience interaction and engagement.

Within discussions of the work, installations and public artwork have a more divided focus. The narratives build threads between theatre and game traditions. They feature experimentation and discuss charitable aims e.g. encouraging people to work together, while being funded by banks and large organisations.

Due to the more varied approaches within public art and sound installations which build on schools of art, audio composition and games, there is not a definitive history being developed specifically for sound but practices are more discussed within other contexts including company mission statement publications.

Audience types

Considering interactive sound installations in a public art concept, Claire Bishop drew links in the current approaches within a UK context with New Labour 'as a form of soft social engineering', while 'the US context, with its near total absence of public funding, has a fundamentally different relationship to the question of art's instrumentalisation'.[57] This consideration suggests that the difference in politics and funding have changed the approaches to public art. I would suggest that this extends past the public art framing to all access to art and museums, where installations may be housed. In the UK, under New Labour, DCMS sponsored museums and galleries became free to the public, increasing the number of annual visitors.[58] This is still a relatively unusual initiative globally, but does increase the access to sound installations not just in public spaces, but in more traditional museum and gallery spaces.

Hesmondhalgh and Pratt discuss the link between histories of cultural policy and the cultural industries, particularly drawing the links between the de-industrialisation of cities in the UK, with the rise of cultural policies and cultural hubs.[59] This is then widened out to the global use of cultural policy in post-industrial settings in countries such as France, Australia, Canada and New Zealand. This narrative particularly looks at the role of class within the creation of cultural hubs which can be seen across narratives. It is important within the creation of public art that the work is not limited to those who can afford art. It creates a narrative of art being central to everyday life, particularly in post-industrial communities. There is a sense that cultural industries and policies should lead to a democratisation of art and culture.

Public policy and access to public art and museums has an impact on audiences who will be experiencing a work. In a culture with more open access to public art, the artist will need to adapt their process to their potential audience, or may make a choice to hold true to the artist's aims. Either way, this has to be a conscious choice that the artist makes.

While the rise of recent approaches to public art and sound installations can be linked to cultural policies, large scale public art and sound installations continue to be a growing part of arts culture in many countries that seems to align to growing acceptance of personal creativity as discussed by Hesmondhalgh and Pratt.[60]

Since this rise in personal creativity in the 90s, there has been a further rise in large-scale experience art dominated by companies, over individuals. Organisations such as Punchdrunk and Secret Cinema stage large scale arts events and Artichoke stages regular large-scale festivals in UK cities. There are an estimated 2106 immersive technology companies in the UK with focuses on virtual or live definitions of immersive.[61] These form part of the 'experience economy' as defined by Pine and Gilmore.[62]

Pine and Gilmore link the risk in company-led experiences and the experience economy to Disney and their theme parks creating a complete experience and bringing additional value to other media properties.[63] While this has created a public appetite for experiences and can still be seen as an important style of experience, this framing links to existing intellectual property and not to new creations. The more recent rise in public experiences, installations and event art seems to coincide with the advent of home streaming, with BBC's iPlayer and Netflix's home streaming platform both launching in 2007, and increased access to good quality technology within the home. If a person wants to watch a film, they can access it from On Demand streaming services and watch it on a large, good quality screen with surround sound at home. This shifts an audience's need to attend cinema events or even some live theatre events. Installations and immersive events such as Secret Cinema offer the audience a community experience that they cannot get at home. Pine and Gilmore link this to our less industrial and agrarian economies allowing for free time and disposable income for experiences.[64]

Interestingly, post-pandemic these shared public experiences appear to have recovered relatively well post-pandemic as demonstrated by Punchdrunk's success with *Burnt City* in 2023, their largest show in scale and their longest run, as it seems that people looked to return to in person events and shared experiences.

Individual artists or companies

Sound art and public sound installations have a fundamental difference in the role of the individual versus the collective, dependent on the location of the art and the potential role of the work.

Representing the academy and gallery approach, the Tate's guide to sound art lists individual artists linked to experimental sound and music including Russolo, Cage, Susan Phillipsz and Bill Fontana.[65] MoMA take a similar approach in their article reviewing the history of sound exhibits, specifically mentioning their first exhibition of sound art in 1979 curated by Barbara London.[66] This links sound art to practices in experimental sound and music.

Within public, interactive sound installations many commissions and events are developed by arts organisations or studio systems rather than individual artists linking to their public roles. These companies can act as a collective, as a commissioning body or with the company name in place of an individual artist. For example, in a curatorial capacity the UK studio Artichoke runs events such as Lumiere, a public art festival featuring a range of artists,[67] while Daily tous les jours and Marshmallow Laser Feast are companies that produce individual pieces of art where the company name acts as the artist.

This is an interesting contrast to sound art or experimental sound and music as it speaks to the ownership of the art which is supported by Bishop's assertion that public art works 'against dominant market imperatives by diffusing single authorship into collaborative activities'.[68] If an individual artist's name is used in a public piece of art then they become the focus of the work and the space. If a collective or a company name is used then an individual is not the focus and can be seen more to be 'serving' the community where the work is being developed. This can be contrasted with the collective aims outlined within experimental interactive sound and music, where the individuals are still a focus of the history even when claiming a more collaborative focus. However, these are still companies, and this collective approach has more of a capitalist framing, there are profitable bids at play and working within a company allows an organisation to expand a workforce within the needs of the art being created without having an impact on the branding and development of the piece.

Collaborative practices

Similar to the experimental sound and music aims outlined by CCRMA and IRCAM, due to the nature of the studio approach in many pieces of interactive sound and music, interactive sound installations tend to include the work of cross disciplinary teams. Marshmallow Laser Feast highlights 'collaboration with artists, scientists, musicians, poets, programmers, engineers and many more makers' in their artist statement,[69] while Daily tous les jours' founders represent diverse skillsets that can be seen across the company as highlighted in the biographies housed on their websites. Mouna Andraos has a background in interactive telecommunications and has worked on interactivity, electronics, and craft,[70] and Melissa Mongiat holds an MA in Creative Practice for Narrative Environments and has worked for a number of large arts organisations in the UK.[71]

The company format that is used in many interactive installations allows for this cross-discipline collaboration to be fully embedded in practices in a way that was encouraged in the exploratory environments such as CCRMA and IRCAM and is required to work on the scale required for large interactive sound installations.

Similarly, companies working on sound installations in public contexts include a commitment to community collaboration in their approaches. For example, Artichoke have a page on 'Participation' on their website including their aims about community engagement, teacher development and under-represented artists and communities,[72] while Random International includes a commitment to 'public co-creation'.[73] The wording of these commitments are important to consider with the community aims of the work as participation can potentially be considered in a more charitable context, as communities are allowed, or enabled, to participate in the work while co-creation would suggest a more equal engagement with the work. This can link to Bishop's writing about the active/passive binary and connotations about class.[74]

The importance of public spaces

A number of companies emphasise the importance of public space within their work, linking this to widening participation aims, ensuring that their art is available to as wide a range of people as possible.

For example, Artichoke, formed in 2005, states on its mission page:

> Artichoke works with artists to invade our public spaces ... Making extraordinary art accessible to all is at the heart of everything we do ... The company has continued to with artists ... with the aim of producing unique, large-scale experiences that appeal to the widest possible audience. We don't believe the arts should only take place behind the closed doors of theatres, concert halls or galleries. Instead, you will find our events in the street, public squares, along the coast or in the countryside.[75]

Canadian company Daily tous les jours specifically sets out an aim addressing public space:

> We believe collective experiences can transform our relationships with our neighbors and our environments. Ultimately, they help connect people and enable change within society.[76]

Umbrellium states their company aim as 'dedicated to transforming urban environments and getting communities meaningfully involved'.[77]

This suggests an importance to companies to situate their work within communities and be of service to communities, suggesting work within a charity framing and focus.

Ownership of public space has an impact on interactive sound installation both in form and in the invitations to collaborate and engage with the work. These missions clearly state that the general public is the proposed audience for their work, situating them away from a gallery setting and aligning them to Bishop's assertion that 'instead of supplying the market with commodities, participatory art is perceived to channel art's symbolic capital towards constructive change'.[78] This will lead to considerations around accessibility in how interaction is designed into a piece, including choices around interaction types and interfaces.[79]

Additionally, the creator needs to consider the site-specific requirements of a space where the creator can either respond directly to the context of the space, in a site-specific context; work in a way that responds to the space in a site sympathetic framing or deliberately work against the requirements of the space, highlighting contrasts or contradictions between the work and the space.

Key themes in public art and sound installations

Access to art impacts the audience types and focus of the work. In a country with more access to public art and galleries, the artist or companies working on a project may need to consider the wider accessibility and impact of their work.

Individual artists represent the academy while studios are more commonly used with public art focuses. This represents the different focuses of the art with company names, allowing for the community to be a larger focus in the work.

Collaborative practices are central to the processes for interactive sound installations. This can be between artists on the project but also with communities with community consultation and development sitting at the centre of work on public art and sound installations. This can also be extended to the role of audience as collaborators and co-creators.

The role of public space can be a consideration for companies working in interactive sound installations and can change based on the context of the work.

Conclusion

Reviewing the histories and stories being told within different areas of interactive sound and music can provide an insight into the aesthetic priorities of each field. While these fields are still in development the stories are still being established and developed through oral histories and retelling of narratives, however we can see some elements being solidified across the more cohesive field of sound and music in games and the more established practices in experimental computer music.

While games, experimental sound and music, public art and sound installations have different artistic aims and some differences in the intended audiences there are some similarities in approach throughout the fields which can be demonstrated through some of the key themes within the histories. This can help us draw links between different sensibilities and approaches in each area.

Centring the audience

The role of the audience is a consideration that links public art and interactive sound installations with game audio. Both address these in different ways. In games, this approach looks at the practical approaches of game mechanics, considering what would further immerse a player within a game and enhance their understanding of gameplay. In public art and sound installations the considerations are more about the role of the audience as co-creator, with audiences having more creative participation in a work, although this can move from participants to co-creators in a work.

In experimental sound and music the audience is little mentioned within narratives as the histories favour approaches that centre the creators, composers and technologists, making innovation a bigger focus than audience.

Collaborators vs individual creators

All three areas focus on the role of collaboration within their histories and discuss the importance of collaborative practices. In games and experimental computer music, this discussion still centres individual genius (despite the history of Bell Laboratories' Experiments in Art and Technologies emphasising that there was no focus on genius). Interactive sound installations do make a distinction between individuals in a gallery or sound art context and companies being used in a public art context. This focus on collaboration across all fields reflects the inbuilt interdisciplinarity of interactive sound and music where technology needs to sit alongside artistic concerns and be developed on an equal standing, where neither the technology or the sound takes precedent.

Roles of industry

All areas of interactive sound and music have links to industry practices within their history of approaches. In games this is the most clearly linked due to the nature of the product being created, while in experimental interactive music this is both a feature through collaborative practices and a tension that sits within the field where profitable technology helps fund further innovation but sits in opposition to open-source, knowledge exchange approaches that sit within the academic communities that participate in the work.

In interactive sound installations and public art, often the creators working as company represent industry within the projects, replicating industry structures and approaches in order to develop artistic products.

Happy accidents in innovation

Alongside the role of the 'genius' in games and experimental sound and music is the repeated concept of 'happy accidents', where technological discoveries are downplayed as being accidental or by-products of other discoveries. This suggests an attachment to innovation seeming innate to the fields or the benefit of experimentation within the development of technology and music.

While these similarities exist within the narratives there are specific differences in the roles of training within the interdisciplinary practices in each field. Experimental interactive sound and music specifically links to schools of composition through practitioners such as Pierre Boulez, which impacts the structural and timbral work being created while game audio prefers discussing technological backgrounds of creators working within the fields.

As the fields using interactive sound and music are still developing, these stories and values will keep developing and being solidified. Each retelling of a history will bring a value more centrally into the field, influencing the approaches as we move forward. There is a potential that in the future we will move to a more solidified musical canon for interactive sound and music. A mindful approach towards the histories as they are being developed will ensure that the canon is as wide-ranging and as inclusive as possible.

Reading group questions

1   If histories point to the aesthetic values in a field, who decides the histories and the stories being emphasised? What is the impact of these choices?

2   Why throughout the histories are discoveries discussed as accidental discoveries? What is the benefit to the field?

3   Is there a conflict between audience-centred approaches and composer-centred approaches? What are these and what impacts do they have on the work?

4   How does a technology-driven approach change the sound?

5   How does an audience-centred approach change the sound and the work being created?

6   Should open-source goals be balanced with more commercial aims? How can this be achieved?

7   How does a commercial structure impact the work being created by public art companies? Is there an impact to having a corporate creator over an individual artist?

8   What is the impact of featuring individual 'geniuses' as a central focus to a history?

9 What role does training play in each history and how does this effect the work
   being created?
10 Once a history or canon has been solidified, can this be redeveloped to centre
   different stories? How can this be achieved?

Notes

1 Michel Chion, *Audio-Vision: Sound on Screen*, 2nd edn (Columbia University Press, 1990).
2 Christopher Small, *Musicking: The Meanings of Performance and Listening* (Wesleyan University Press, 1998).
3 Nick Yee, 'Gaming Motivations Group Into 3 High-Level Clusters', *Quantic Foundry*, 2015 <https://quanticfoundry.com/2015/12/21/map-of-gaming-motivations/> [accessed 12 March 2024].
4 Karen Collins, *Game Sound: An Introduction to the History, Theory, and Practice of Video Game Music and Sound Design* (The MIT Press, 2008), doi:10.7551/mitpress/7909.001.0001.
5 These semiotic associations will be further discussed in chapter 3
6 Megan Lim, Ashley Brown, and Ari Shapiro, 'Pong Was Released by Atari 50 Years Ago', *NPR*, 2 December 2022, section Humor & Fun <https://www.npr.org/2022/12/02/1140441610/pong-was-released-by-atari-50-years-ago> [accessed 9 March 2024].
7 Carol Truxal, 'The Secrets of Space Invaders - IEEE Spectrum' <https://spectrum.ieee.org/space-invaders> [accessed 9 March 2024].
8 'Shooting from the Hip: An Interview with Hip Tanaka' <https://www.gamedeveloper.com/audio/shooting-from-the-hip-an-interview-with-hip-tanaka> [accessed 9 March 2024].
9 Collins, p. 39.
10 Collins, p. 40.
11 *Call of Duty: Modern Warfare 2*, 2009.
12 Mike Snider, 'Interview: "Modern Warfare 2" Composer Hans Zimmer', *USA Today*, 2009 <http://content.usatoday.com/communities/gamehunters/post/2009/11/qa-with-modern-warfare-2-composer-hans-zimmer/1?csp=34> [accessed 25 March 2024].
13 Collins, p. 35.
14 This will be discussed in depth in later chapters.
15 Andrew J. Nelson, *The Sound of Innovation: Stanford and the Computer Music Revolution* (The MIT Press, 2015), chap. 3.
16 Center for Computer Research in Music and Acoustics, *Brief History*, p. 2 <https://ccrma.stanford.edu/~aj/archives/docs/all/646.pdf> [accessed 9 March 2024].
17 IRCAM, 'History' <https://www.ircam.fr/lircam/historique> [accessed 21 March 2024].
18 Boulez p. 8
19 Nelson, pp. 23–25.
20 Julie Martin, 'A Brief History of Experiments in Art and Technology', *IEEE Potentials*, 34.6 (2015), pp. 13–19 (p. 14), doi:10.1109/MPOT.2015.2443897.
21 Nelson, p. 31.
22 Georgina Born, *Rationalizing Culture: IRCAM, Boulez, and the Institutionalization of the Musican Avant-Garde* (University of California Press, 1995), pp. 234–35.
23 IRCAM, 'Research' <https://www.ircam.fr/recherche> [accessed 15 March 2024].
24 IRCAM, 'Creation' <https://www.ircam.fr/creation> [accessed 15 March 2024].
25 Nelson, p. 5.

26 John Beck and Ryan Bishop, *Technocrats of the Imagination: Art, Technology and the Military Industrial Avant-Garde* (Duke University Press, 2020), p. 4.

27 Pierre Boulez, *Music Lessons: The Collège de France Lectures*, ed. by Jonathan Dunsby, Jonathan Goldman, and Arnold Whittall, trans. by Jonathan Dunsby, Jonathan Goldman, and Arnold Whittall (Faber & Faber, 2018), p. 10.

28 Boulez, p. 11.

29 Boulez, p. 13.the

30 Beck and Bishop, p. 5.

31 Born, p. 66.

32 Nelson, chap. 4.

33 Born, p. 138.

34 Beck and Bishop, p. 2.

35 Conventional DAWs can be seen to do this through their default template with tempo, time signature and key signature already set. The composer has to actively choose not to use the pre-set formats.

36 Miller Puckette, 'Max at Seventeen', *Computer Music Journal*, 26.4 (2002), pp. 31–43 (p. 31).

37 Puckette, p. 41.

38 Eric Lyon et al., 'Dartmouth Symposium on the Future of Computer Music Software: A Panel Discussion', *Computer Music Journal*, 26.4 (2002), pp. 13–30 (p. 20).

39 Boulez, p. 9

40 Center for Computer Research in Music and Acoustics, p. 2.

41 'Nokia Bell Labs History', *Nokia Bell Labs*, 2020 <https://www.bell-labs.com/about/history/#gref> [accessed 9 March 2024].

42 IRCAM, 'History'.

43 Center for Computer Research in Music and Acoustics, p. 3.

44 Born, p. 67.

45 Born, pp. 68–69.

46 Niels J. Reimers and Larry Horton, 'Reimers, Niels J.' <https://purl.stanford.edu/sk314qb9448> [accessed 22 March 2024].

47 Nelson, p. 42.

48 YLEM: Artists Using Science and Technology, *YLEM Journal Volume 25 Issue 6 and 8*, 2005, p. 6 <http://archive.org/details/ylem-journal-v25i06_08> [accessed 22 March 2024].

49 John M. Chowning, 'Method of Synthesizing a Musical Sound', 1977 <https://patents.google.com/patent/US4018121A/en> [accessed 16 July 2024].

50 Reimers and Horton, pp. 52–54.

51 Nelson, p. 86.

52 Born, p. 184.

53 The software name Max was taken after Max Mathews reflecting his central role in the development of computer music.

54 Darwin Grosse, 'Art + Music + Technology: Podcast 090: Miller Puckette' <https://artmusictech.libsyn.com/podcast-090-miller-puckette> [accessed 22 March 2024].

55 Lyon and others, p. 14.

56 Puckette, p. 41.

57 Claire Bishop, *Artificial Hells: Participatory Art and the Politics of Spectatorship* (Verso, 2012), p. 5.

58 DCMS, 'Ten Years of Free Museums', *GOV.UK* <https://www.gov.uk/government/news/ten-years-of-free-museums> [accessed 23 March 2024].

59 David Hesmondhalgh and Andy C. Pratt, 'Cultural Industries and Cultural Policy', *International Journal of Cultural Policy*, 11.1 (2005), pp. 1–13 (p. 3), doi:10.1080/10286630500067598.

60 Hesmondhalgh and Pratt, p. 4.

61 'Innovate UK Immersive Tech Network → The 2022 UK Immersive Economy Report' <https://iuk.immersivetechnetwork.org/resources/the-2022-uk-immersive-economy-report/> [accessed 24 March 2024].

62 B. Joseph Pine II and James H. Gilmore, *The Experience Economy* (Harvard Business Review Press, 2019).

63 Pine II and Gilmore, p. 18.

64 Pine II and Gilmore, p. 22.

65 Tate, 'Sound Art' <https://www.tate.org.uk/art/art-terms/s/sound-art> [accessed 23 March 2024].

66 The Museum of Modern Art, 'Resonant Frequencies: Sound at MoMA | Magazine | MoMA', <https://www.moma.org/magazine/articles/877> [accessed 23 March 2024].

67 Artichoke, 'Lumiere | Light Art Festival | November 2023' <https://www.lumiere-festival.com/> [accessed 15 March 2024].

68 Bishop, p. 12.

69 Marshmallow Laser Feast, 'Features' <https://marshmallowlaserfeast.com/> [accessed 24 March 2024].

70 Daily tous les jours, 'Mouna Andraos | Daily Tous Les Jours' <https://www.dailytouslesjours.com/en/about/team/mouna-andraos> [accessed 24 March 2024].

71 Daily tous les jours, 'Melissa Mongiat | Daily Tous Les Jours' <https://www.dailytouslesjours.com/en/about/team/melissa-mongiat> [accessed 24 March 2024].

72 Artichoke, 'Participation - Artichoke' <https://www.artichoke.uk.com/participation/> [accessed 24 March 2024].

73 RANDOM INTERNATIONAL, 'Biography' <https://www.random-international.com/biography> [accessed 24 March 2024].

74 The effectiveness of collaboration and audience roles in a collaborative work will be discussed further in chapters 2 and 3.

75 Artichoke, 'Our Mission - Artichoke' <https://www.artichoke.uk.com/our-mission/> [accessed 23 March 2024].

76 Daily tous les jours, 'Mission | Daily Tous Les Jours' <https://www.dailytouslesjours.com/en/mission> [accessed 24 March 2024].

77 Umbrellium, 'About Us' <https://umbrellium.co.uk/about/> [accessed 24 March 2024].

78 Bishop, p. 13.

79 Interface design and encouraging audiences to interact will be discussed further in chapter 4.

References

Artichoke, 'Lumiere | Light Art Festival | November 2023', *Lumiere Festival* <https://www.lumiere-festival.com/> [accessed 15 March 2024].

Artichoke, 'Our Mission - Artichoke' <https://www.artichoke.uk.com/our-mission/> [accessed 23 March 2024].

Artichoke, 'Participation - Artichoke' <https://www.artichoke.uk.com/participation/> [accessed 24 March 2024].

Beck, John and Ryan Bishop, *Technocrats of the Imagination: Art, Technology and the Military Industrial Avant-Garde* (Duke University Press, 2020).

Bishop, Claire, *Artificial Hells: Participatory Art and the Politics of Spectatorship* (Verso, 2012).

Born, Georgina, *Rationalizing Culture: IRCAM, Boulez, and the Institutionalization of the Musican Avant-Garde* (University of California Press, 1995).

Boulez, Pierre, *Music Lessons: The Collège de France Lectures*, ed. by Jonathan Dunsby, Jonathan Goldman and Arnold Whittall, trans. by Jonathan Dunsby, Jonathan Goldman and Arnold Whittall (Faber & Faber, 2018).

Brandon, Alexander, 'Shooting from the Hip: An Interview with Hip Tanaka' <https://www.gamedeveloper.com/audio/shooting-from-the-hip-an-interview-with-hip-tanaka> [accessed 9 March 2024].

Center for Computer Research in Music and Acoustics, 'Brief History' <https://ccrma.stanford.edu/~aj/archives/docs/all/646.pdf> [accessed 9 March 2024].

Chion, Michel, *Audio-Vision: Sound on Screen*, 2nd edn (Columbia University Press, 1990).

Chowning, John M., 'Method of Synthesizing a Musical Sound', 1977 <https://patents.google.com/patent/US4018121A/en> [accessed 16 July 2024].

Collins, Karen, *Game Sound: An Introduction to the History, Theory, and Practice of Video Game Music and Sound Design* (The MIT Press, 2008), doi:10.7551/mitpress/7909.001.0001.

Daily tous les jours, 'Melissa Mongiat | Daily Tous Les Jours' <https://www.dailytouslesjours.com/en/about/team/melissa-mongiat> [accessed 24 March 2024].

Daily tous les jours, 'Mission | Daily Tous Les Jours' <https://www.dailytouslesjours.com/en/mission> [accessed 24 March 2024].

Daily tous les jours, 'Mouna Andraos | Daily Tous Les Jours' <https://www.dailytouslesjours.com/en/about/team/mouna-andraos> [accessed 24 March 2024].

DCMS, 'Ten Years of Free Museums', GOV.UK <https://www.gov.uk/government/news/ten-years-of-free-museums> [accessed 23 March 2024].

Grosse, Darwin, 'Art + Music + Technology: Podcast 090: Miller Puckette' <https://artmusictech.libsyn.com/podcast-090-miller-puckette> [accessed 22 March 2024].

Hesmondhalgh, David and Andy C.Pratt, 'Cultural Industries and Cultural Policy', *International Journal of Cultural Policy*, 11. 1 (2005), pp. 1–13, doi:10.1080/10286630500067598.

'Innovate UK Immersive Tech Network → The 2022 UK Immersive Economy Report' <https://iuk.immersivetechnetwork.org/resources/the-2022-uk-immersive-economy-report/> [accessed 24 March 2024].

IRCAM, 'Creation' <https://www.ircam.fr/creation> [accessed 15 March 2024].

IRCAM, 'History' <https://www.ircam.fr/lircam/historique> [accessed 21 March 2024].

IRCAM, 'Research' <https://www.ircam.fr/recherche> [accessed 15 March 2024].

Lim, Megan, Ashley Brown and Ari Shapiro, 'Pong Was Released by Atari 50 Years Ago', *NPR*, 2 December2022 <https://www.npr.org/2022/12/02/1140441610/pong-was-released-by-atari-50-years-ago> [accessed 9 March 2024].

Lyon, Eric, Max V. Mathews, James McCartney, David Zicarelli, Barry Vercoe, Gareth Loy, Miller Puckette, 'Dartmouth Symposium on the Future of Computer Music Software: A Panel Discussion', *Computer Music Journal*, 26. 4 (2002), pp. 13–30.

Marshmallow Laser Feast, 'Features' <https://marshmallowlaserfeast.com/> [accessed 24 March 2024].

Martin, Julie, 'A Brief History of Experiments in Art and Technology', *IEEE Potentials*, 34. 6 (2015), pp. 13–19, doi:10.1109/MPOT.2015.2443897.

The Museum of Modern Art, 'Resonant Frequencies: Sound at MoMA | Magazine | MoMA' <https://www.moma.org/magazine/articles/877> [accessed 23 March 2024].

Nelson, Andrew J., *The Sound of Innovation: Stanford and the Computer Music Revolution* (The MIT Press, 2015).

Nokia Bell Labs, 'Nokia Bell Labs History', 2020 <https://www.bell-labs.com/about/history/#gref> [accessed 9 March 2024].

PineII, B. Joseph, and James H. Gilmore, *The Experience Economy* (Harvard Business Review Press, 2019).

Puckette, Miller, 'Max at Seventeen', *Computer Music Journal*, 26. 4 (2002), pp. 31–43.

RANDOM INTERNATIONAL, 'Biography' <https://www.random-international.com/biography> [accessed 24 March 2024].

Reimers, Niels J. and Larry Horton, 'Reimers, Niels J.' <https://purl.stanford.edu/sk314qb9448> [accessed 22 March 2024].

Small, Christopher, *Musicking: The Meanings of Performance and Listening* (Wesleyan University Press, 1998).

Snider, Mike, 'Interview: "Modern Warfare 2" Composer Hans Zimmer', *USA Today*, 2009 <http://content.usatoday.com/communities/gamehunters/post/2009/11/qa-with-modern-warfare-2-composer-hans-zimmer/1?csp=34> [accessed 25 March 2024].

Tate, 'Sound Art' <https://www.tate.org.uk/art/art-terms/s/sound-art> [accessed 23 March 2024].

Truxal, Carol, 'The Secrets of Space Invaders - IEEE Spectrum' <https://spectrum.ieee.org/space-invaders> [accessed 9 March 2024].

Umbrellium, 'About Us' <https://umbrellium.co.uk/about/> [accessed 24 March 2024].

Yee, Nick, 'Gaming Motivations Group Into 3 High-Level Clusters', *Quantic Foundry*, 2015 <https://quanticfoundry.com/2015/12/21/map-of-gaming-motivations/> [accessed 12 March 2024].

YLEM: Artists Using Science and Technology, *YLEM Journal*, 25. 6 & 8, 2005 <http://archive.org/details/ylem-journal-v25i06_08> [accessed 22 March 2024].

Media and art examples

Infinity Ward. *Call of Duty: Modern Warfare 2* (Activision, 2009).

Chapter 1

Is it actually interactive and does it actually matter?

When experiencing interactive sound and music there is an implied contract in place between the creator and the audience that the interaction within a piece is genuine and that their actions are having an impact on the work. However, there may be cases where work is advertised as interactive where the actual interaction and impact on the work is minimal or the work is not interactive at all.

However, this may not impact the audience experience of the work who perceive the piece to be interactive.

This chapter will address what makes a work interactive and whether this impacts an audience's experience of the musical work. It will address the definitions of interactive sound and music and then apply these to examples across a wide range of approaches to interactive audio.

Beyond pressing play: defining interaction

In order to determine if a piece of work is authentically interactive we really need to clarify how we are defining interactive sound and music. You would think this would be simple to answer but like many areas, such as defining work as immersive, this is up for debate and is dependent on context provided in different fields.

As a very simple response, when I am asked what I mean I usually define interactive sound and music as 'you do something and the sound reacts'. It's simple and succinct and it doesn't hold up the conversation for too long. For the purposes of this chapter, it doesn't actually provide the necessary specificity for us to understand the authenticity of interactive experiences.

Often when discussing interactive sound and music, literature often defaults to interactive technology in a performance context based on human-computer interaction. This approach focuses on the relationships between humans and technology and how this can enhance a practice. This builds on the history of experimental sound and music, and research preoccupations of this field as outlined in the introduction. For example, an early example of score-following software developed at IRCAM is demonstrated in Boulez's Anthèmes II.[1] In this work, the software listens to the performer and adjusts to their performance

DOI: 10.4324/9781003344148-2

pace and mannerisms in a way that a human accompanist would. The software is interacting with the performer. This field can also move to controllerism, where MIDI objects or game controllers are used as an interactive tool, providing a particular benefit for the performer composer over the fine details of their sound in performance.[2] A famous example of this type of human–computer interaction can be seen through Imogen Heap's MiMu glove project which attempts to bring more gesture control and performativity to the use of controllers and interfaces.[3]

However, I would argue that human–computer interaction in a performance context where the interaction with technology is instigated by a composer is an extension of instrument building rather than interactive practice. The technology is allowing the performer to have greater flexibility in their work, but they are still acting as a mediator with the audience taking on traditional roles by observing the performance practice.

Katja Kwastek provides the following qualifier when defining interactivity that 'interactive art places the action of the recipient at the heart of its aesthetics.'[4] This is particularly applicable when looking at interaction in a games or public art context. The audience is central to the approach. So a central tenet to the definitions of interaction for the purposes of this chapter is that the sound is influenced by someone other than the composer or performer, taking the music out of a traditional performance landscape.

Another limitation with the term human-computer interaction is that it really confines the interaction to a technology-based approach – meaning that you can miss out on some very simple forms of interaction, e.g. a bell attached to a help desk is a pretty satisfying form of interaction where the person is immediately responsible for an interactive response.

There's also a question of does a human need to interact with a work to make it interactive? For example, Céleste Boursier-Mougenot's 2010 *From Here to Ear* at the Barbican's Curve Gallery,[5] where the piece consisted of an aviary built from guitars where finches could land on the instruments, relied on birds interacting with their surroundings. This could be considered interactive work (unless we consider the birds to be performers). Equally, you could design a work that was intended to be interacted with by machines or live data feeds. For example, Sun Yuan and Peng Yu's installation *Can't Help Myself*[6] features a robot that senses when the fluid in the space has moved too far away and then acts to return it to its original position. This piece could be considered an interactive work as the robot is using sensors to influence its decisions so would still be as unpredictable as human interaction. Similarly, a piece that uses live datasets, such as Hatnote's *Listen to Wikipedia* that generates sound based on Wikipedia edits,[7] is also reacting in real time to an outside stimulus. Although this does bring into question generative composition rather than pre-composed music and how this sits in an interactive context. As this is a real-time generation of audio that is dependent and reactive to live data, rather than a pre-composed work or a piece where the algorithmic input remains consistent, this can be considered an interactive piece of work.

The consistency of interaction within a piece is also an important consideration when we are defining interactive sound and music. You could argue that
pressing a button and having audio play is technically interactive, making all
streaming platforms, record players etc. pieces of interactive sound and music.
The sound cannot start without the listener's action and it is therefore entirely
dependent on the listener's action. However, once the audio has started this
interaction stops. The listener's input is no longer required to make the audio
continue to work or develop. To be truly interactive, the process needs move
beyond pressing play to be more consistent and reliant on real-time audience or
data input. Further to this, any action from the audience or data must have a
tangible impact on the work being created.

With these conditions in mind, this chapter, and book, is written with the
following requirements for audio to be considered interactive:

- The sound and music is influenced by someone who is not the performer or
  composer. This can include, but is not limited to, audience members,
  animal or robot participants or live datasets
- The input from the outside influence must be ongoing and have a continuous
  effect on the work being created. It cannot be a one-time instigating action
- The action from the stimulus influences the outcome of the resulting audio
- Interactivity within a work does not need to be technological or limited to
  a digital domain and can include more mechanical interactivity such as use
  of bells

A scale of interactivity

A way to consider the level of engagement in interactive sound and music is in
the context of Sherry Arnstein's Citizen Ladder of Citizen Participation.[8] The
Ladder of Citizen Participation is an architectural theory that describes how
empowered a community is within a project from levels of non-participation
which includes manipulation and therapy approaches; tokenism including
informing and consulting and citizen power including partnership, delegated
power and building to citizen control. This is a particularly useful tool when
considering public art projects and can be applied to the level of interactivity
from traditional audience roles through to complete audience control.

The amount that each work addresses the requirements for interactivity,
aligned to Arnstein's Ladder of Citizen Participation, can vary based on the
style of work, composer intentions and the expectations for the audience. This
can be developed to a scale of interactivity.

Perceived interaction is where the audience feel that they are interacting with
the work but their actions are not influencing the final sound or outcomes of
the project. For example, in Radiohead's 2014 PolyFauna app album the user is
able to explore landscapes and soundscapes on a smartphone or tablet.[9] The
exploration by the user does not affect the final outcome of the sound being

played and, at times, cinematics are used to move a scene on. By limiting the scope of the interaction within the work the composer effectively prioritises the completeness of their musical sound over the audience experience but the audience feel that they have had an interactive experience through the work.

Nominal interaction allows the audience to have small amounts of interaction with the music without impacting the overall soundscape or composition. For example, in the rhythm game *Sayonara Wild Hearts*[10] players are moving through a pre-composed soundtrack. If they miss an action that syncs to the rhythm the game brings them back to the last game checkpoint where the track can restart. This allows the audience some control over their experience but within very set parameters so the composer retains complete control over the output of the work.

'On the rails' interaction allows some interaction that does not change the overall structure of the sound and music as players or audience members are still expected to move to certain narrative or musical points within the game. For example, in the game *Journey*,[11] while you are moving through a space and meeting other characters, the actions that you take on screen do not impact the outcomes of the game, just how you travel through the narrative. Within this approach the player or audience have explorative control over their experience in the work and are able to pace their approach to the narrative, but still reach all points within the narrative as the composer or creative intends.

Adaptive audio changes based on the player or audience actions but does still follow narrative. For example, in AAA games adaptive audio will trigger different audio cues and provide a live mix of different intensities.[12] Players will have increased exploratory control and will have an experience that adapts to how they approach the game, including supporting emotional developments within a narrative, but they will not affect the overall music being created in the space, which will be pre-composed.

Structured generative audio/processes create a lot more freedom in the interactive processes where the music is generated or composed based on the interaction, which could be audience/player actions, external data (as in *Listen to Wikipedia*) or other external focuses. For example, in the game *Ape Out*, the drum soundtrack generates based on player behaviours with more aggressive behaviours resulting in a louder, more percussive soundtrack.[13]

Free interaction with creative agency allows the audience to have full control over their experience within the sound world. This is a sandbox experience of interaction where the composer has exerted minimal control over the compositional output, instead creating the conditions for interaction to take place. For example, the Reactable provided all the building blocks for people to develop music through the physical interface, but did not place any restrictions on the audience.[14]

The difference between these modes could also be seen as the difference between attending, participating and collaborating.

- When attending a work, traditional audience roles are upheld. While the audience may be able to explore the space they are not impacting the sound being created
- When participating, the composer invites the audience to engage with the work but within conditions and approaches set by the composer
- When collaborating the audience member is working with the composer. They have agency to work with and contribute to the music being created

Table 1.1 Scale of interaction linked to Arnstein's Ladder of Citizen Participation

Minimal interaction					*Completely interactive*
Non-participation		Tokenism			Citizen Power
Perceived interaction	Nominal interaction	'On the rails' interaction	Adaptive audio	Structured generative audio/ processes	Free interaction with creative agency
Attendance		Participation			Collaboration

Control vs collaboration

The scale of interaction represents an artistic choice related to interaction and the composer's intention related to how much creative control they want to exert over the work in favour of audience interaction. In traditional composition practices the composer is seen to have complete control over the work.[15]

Within a lot of classical compositional training the composer is taught to control every element of the work. This starts from understanding how to construct harmony and counterpoint to building detailed scores that include all the elements of notation and expression to support the performer to interpret the work in the way intended by the composer.

In order to create fully interactive work the composer must release some control over the sound and structure of the pieces. By including interaction within a piece of music, it is vital that the interaction actually contributes to the audience's experience and the sounds that they hear otherwise the interaction may just seem like a gimmick to audience members. To relate this back to Arnstein's Ladder of Citizen Participation, to move the audience from traditional roles where they are informed or consulted up to the higher levels of participation of delegated power and citizen control the composer needs to change their own role in the project.[16] As Claire Bishop summarises in *Artificial Hells*, 'the fact that the Ladder of Participation culminates in "citizen control" is worth recalling here. At a certain point, art has to hand over to other institutions if social change is to be achieved'.[17]

While Bishop's conclusion has been written with art that has a political objective in mind, it can be applied to interactive sound and music. In order for the composer to achieve their full artistic aim, control needs to be handed over to the audience; however, how much control is handed over is the choice of the composer and does fundamentally change the power structures between the composer and the audience or player. The composer is no longer the only person with creative control over the work.

Giving up more control to the audience creates a question of ownership over the final work and the sound produced. If the audience are responsible for all the sound created, can the composer fully claim responsibility for all the outcomes of the interactive work?

A similar scenario can be drawn with open improvisatory compositions such as text pieces and graphic scores. In these more open improvisatory pieces, the composer has set the conditions for the work to take place but has not chosen the notes and rhythms being performed. For example, Karlheinz Stockhausen's *Aus den sieben Tagen* is considered by the composer to be 'intuitive music'.[18] These 15 text pieces provide a set of conditions or meditations for performers who improvise based on the conditions set, which are designed to ensure that the performers don't rely on their preferred improvisation approaches. In this case, the composer sets all the conditions and the work cannot exist without them. However, the work only exists in principle until the performers take the instructions and use them to provide the performance. Stockhausen has been known to exercise a large amount of control over his work as a composer, even when working with intuitive music processes. In the case of *Aus Den Sieben Tagen,* while setting the conditions for intuitive or improvisatory practices Stockhausen reportedly became angry when the music produced did not fit his ideal for the work.[19] Stockhausen also viewed himself as owning all the outputs of the work created with the performer acting as a 'radio receiver' for his work.[20]

Stockhausen's perception maintains the traditional hierarchies of the composer even when working with more intuitive or improvisatory processes. However, if the composer is aiming to work towards citizen power within Arnstein's Ladder then this perception of control and ownership needs to shift when discussing the role of the audience, or other instigator of the interaction.

Rather than claiming full ownership over any interactive work where the audience is contributing to the final musical output, the process could be considered to be a long distance collaboration. The composer has provided the tools that help the audience to complete the sound for the work in a way that they see fit.

Reasons for limiting interaction

Some of the reasons why a composer or sound designer may wish to limit interaction in their work or keep it 'on the rails' are related to the practicalities of the work and linked to narrative and artistic intentions.

Very practically speaking, if a sound installation is running for the duration of a gallery opening there may be some technical glitches that stop the work from functioning. Rather than close an exhibit while the work is rebooted it can be better to have a version that appears to be interactive but instead is providing pre-set or recorded responses. This is done with audience experience in mind as it ensures continuity of an installation. However, arguably, some audience members are getting a less 'complete' experience. There is an alternative option of running the installation software on multiple machines so you can switch machines while a glitch is fixed, but that is quite a cost intensive approach.

Interaction can reach a saturation point of people, where the interaction is unable to cope leading to no changes in sound or causing the installation to crash. For example, *Dune*, a sound and light installation by Studio Roosegaarde, features shrub-like objects that react to changes in movement and sound around them, creating a whispering soundscape reacting to the visitors.[21] In 2009, as part of the first Lumiere Durham festival, *Dune* was installed in Durham Cathedral Cloisters. The large number of visitors to the festival meant that, although still an effective installation within the space, there was little change in sound or light. One solution to this is to stagger entry to the installation, to cap the audience numbers and start a queuing system. This is similar to the approach taken by the Barbican for Random International's *Rain Room* in 2013, an installation where falling rain responded to audience members' movements allowing them to walk through a downpour of rain while staying dry.[22] On 2 March 2013, the queue for *Rain Room* stood at 8 hours.[23] However, if made to queue for a long time without any additional artwork to focus on, audience members may lose interest and leave without seeing the installation. To address this saturation point in the interaction a composer may wish to limit the number of options for interaction so that the work can deal with increased numbers without impacting the overall sound while preserving the group experience and reducing queuing times.

Outside of the practical, there are narrative reasons why interaction may be kept to a minimum or be 'on the rails'. For example, the game *Florence*[24] follows the relationship between Florence and her partner, a musician, Krish. The interaction within the music in this game is very much on rails, building to a sense of inevitability in their relationship. There are places where perceived interaction and control in the audio can be an effective narrative choice, particularly if, as with *Florence*, these choices are demonstrating an inevitable outcome or a lack of control for the characters. These sit alongside artistic considerations when considering limiting interaction in a work, a composer or artist may be making a political point through limiting the amount of interaction in a work by showing the limit of the influence that an person can have over the work. In the previous example of *Can't Help Myself*,[25] the robot's actions are severely limited. It cannot move closer to the oil as it is being leaked and it cannot stop the oil from leaking, it can only prolong the inevitable by replacing what it can as it sees the leak.

Musically speaking there may be times when the composer needs to retain some authorial control. Free interaction within an interactive work is developed with an aim that only the conditions for the interaction and the audio is set by the composer, with all other decisions around the work sitting with the audience or instigator of the interaction. If aiming for completely free interaction the composer faces some challenges around the cohesion of the project. Allowing absolutely free interaction means giving up all composer control which can impact the melodic and harmonic sound within the work which can begin to sound unstructured, particularly if a saturation point of interaction is reached.

Creatively, a composer may wish to reduce interaction if they have a very specific sound in mind. For example, if they are aiming for a particular chord pattern or structure within the work. This can include ensuring that the music within the work is coherent and ensuring a consistent audience experience, even when there is some flexibility. For the composer to retain this authorial control, interactive processes can be limited in the number of outcomes that they could produce – for example limiting the types of melodic material so that all the audio is concordant, ensuring that the sound within the installation remains consistent for each audience member within the installation while allowing for their own variations and interpretations of the work. Using the Ladder of Citizen Participation as a framework, this would be classified as 'delegated power'. The audience is not in complete control but makes decisions in lieu of the composer.

Will the audience actually know if it's not interactive?

If a composer chooses not to delegate power with their work while still labelling it as an interactive process, how would an audience be able to tell if they are not interacting with the music?

When an audience engages with an interactive piece of work they enter into an agreement with the composer about the conditions of the work. This includes that their actions are actually impacting the work being produced. If the composer chooses to present a piece of work as being interactive when it is not then they are breaking that social agreement that they have made with the audience. While there could be artistic reasons for doing this, this could greatly erode the trust that the audience has in the composer and impact their engagement with the work.

There are ethical implications to misleading an audience about the apparent amount of interactivity in the work.

However, the question of whether the audience would know if a work is actually interactive is difficult to answer. When attending exhibitions you can observe an audience's behaviour around an interactive piece and see how they are engaging with the work and apparent interactivity. There are a couple of ways to observe whether they believe the work to be interactive.

The first is how many times they repeat an interactive movement, testing to see if their interaction has registered. Similar to when someone is repeatedly pressing the button at a pedestrian crossing, people will continuously press a button until they get an interactive result which they then attribute to the repeated action.[26]

If an audience believes that their interaction is making no difference to the work, they may declare that the exhibit is broken, citing the lack of instant reaction as evidence.

Finally, the time that an audience member spends with an exhibit is usually evidence of the effectiveness of the interaction. If their actions are having a credible effect on the work then they will stay longer in the exhibit, testing the limits of the piece.

Composer/audience obligations with interaction

When a composer sets out to create an interactive piece of work, they enter into a social contract with the audience about the authenticity of the interaction. While the audience or instigator for the interaction in the work is considered a collaborator in the piece, then they should be treated as such with the respect that would be given to a collaborator. This social contract becomes particularly important when working on sound installations in a public art context. Within public art the composer or artist becomes a guest of a community and any social contract would build their trust with the community, centring their role within the work.

These elements of social contracts are established through how the interaction is established within a work. For example, if you look at a theatre company such as Punchdrunk their work is constructed in a similar way. 2023's *The Burnt City* was constructed through a number of interweaving narratives linking Agamemnon and the Fall of Troy. The work was made of two warehouses with multiple rooms connected by a small corridor. This allowed the audience to move freely between the worlds, the way they could in a game. Additionally, they could actively seek out the narrative by following actors in the space or by inspecting environmental objects such as diaries or newspapers. Famously, audience members could be invited into a private space for a one-to-one interaction with characters, acting in a similar way to a side quest or easter egg in a game. In order for this space to be successful, audience members had to be emboldened to interact with the work and had to have the conditions of the interaction established before entering the space, similar to the way that they would in a game.

This was encouraged through creating a set of rules that the audience signed up to in the space, which were outlined verbally before entering the space:

1   Mobile phones are not allowed
2   All audience members must wear a mask at all times
3   You are not allowed to talk
4   You can look at objects but they must be left as found for the next audience members to find

The mask acted as a physical contract with the audience, once it was worn you had the freedom to explore the space, but you had also agreed to the prior rules for behaviour.

In games, the contracts for interaction are established through training levels, where the limits and structures of the game are presented to the player. They outline the controls for interaction but also the amount of freedom within a game. For some games this is a short introduction with different elements being added as needed within gameplay, while in an open world game such as *The Legend of Zelda: Breath of the Wild* the training level will be extensive and provide avenues for free exploration alongside learning mechanics.[27] In games, the genre and style of game is a key component for establishing the contract for interaction. A game such as *Sayonara Wild Hearts*[28] that prioritises the music while limiting interaction establishes that priority within the opening of the game and by self-describing as a 'pop album video game'.[29] This makes it clear to the audience that they are experiencing an album rather than creating the sound. *Ape Out*[30] has a similar album-like structure to the game with each level representing a different style of jazz. The first level establishes a different contract with the player by demonstrating the direct link between the user's actions and the music being created. This contract is, again, outlined in the game's storefront that highlights the 'Dynamic Soundtrack' where you can 'find your rhythm in the chaos as a dynamic soundtrack of drums, cymbals, and decapitations drive the action to the edge of mayhem'.[31]

In a sound installation the contract may require more establishment based on the style of interaction being involved. In a gallery setting this may be established by an instruction panel for the work, or by ushers within the gallery providing an introductory talk or instruction for the work. While this clearly establishes an expectation for the work, it does establish a formality that links it to the gallery environment. In most successful interactions the conditions and social contract within the gallery space are made clear without providing extensive additional instruction. For example, Yayoi Kusama's *Obliteration Room*[32] provides audiences with a sheet of stickers and a blank room where they can cover every surface with dots. With the Obliteration Room example, the contract becomes clearer as the work progresses. For the first people interacting with the space, the white walls are prohibitive. As more people engage with the work and with the space, the interaction becomes more intuitive and clearer to the audience. For this style of interaction where there may be inhibition, the composer or artist can create a supportive environment by starting the interaction in advance or adding some 'plants' to the audience who can demonstrate the interaction.

In a public setting, the contract for interaction can either work with or against the site, providing a site sympathetic approach or a conflicting approach. Companies like Daily tous les jours choose to work with the public understanding of the space, building a contract through commonly known interactions that the public will understand without further explanation such as playing on swings or using stepping stones.[33] Through this a safe environment for interaction is created, allowing for the discovery of musical features within the work.

While these examples all differ in how they approach the concept of the composer/audience obligations within a work, they have two core elements in common that form a social contract with the audience:

1   The work will truly be interactive – while the elements of interaction may differ based on practicalities and narrative functions, there will be interaction at the heart of the work
2   The audience or instigator's actions will impact the work – the changes happening to the sound will be authentic and genuine to the actions undertaken

Authenticity in interaction

The composer's obligations to the audience when creating an interactive piece are based on establishing a sense of authenticity of the creator and, therefore, the work. Authenticity is an often debated concept within wider fields of sound and music, particularly in commercial sound and music. As a concept authenticity can often be considered flawed as an approach, particularly as it is applied often to guitar-led work leading to authenticity being viewed through a particular lens of male whiteness.[34]

Conversations of authenticity extend to performance practices as demonstrated by organisations such as the Orchestra of the Age of Enlightenment,[35] where historical performance techniques and tunings are used to accurately represent the work as it was originally heard.[36]

Authenticity can be hard to pin down. As a framing it points to the work being developed with a sense of honesty to the artist where they are representing their experience through the work. In popular music, this can mean representing their lived experiences through the work, particularly when relating to elements of class or personal storytelling within a piece of work. In classical music, this points to representing the composer's intention or the historic practices represented within the work. In public art, when working with communities, the artist should work to fully represent that community, and feature the voices of the community as collaborators with agency within the work rather than imposing the artist's viewpoint on a community.

In interactive sound and music, the term authenticity can be applied to the construction of the work and the processes of interaction. The critic Walter Benjamin discusses authenticity within the context of original work versus reproductions. Benjamin discusses the 'here and now' of the artwork as representing a difference to experiencing work through a digital reproduction. When discussing the representation of the actor in film, Benjamin states:

...for the first time – and this is the effect of film – the human being is placed in a position where he must operate with his whole living person, while forgoing its aura. For the aura is bound to his presence in the here and now. There is no facsimile of the aura.[37]

While the 'aura' in this statement refers to the sense of being in the room with an actor during a stage performance, this can equally be applied to interactive sound and music. In an interactive environment the aura is the feeling of influencing the sound and music being created. It is a feeling that is hard to fully define, but it is the sense of being in the presence of the work and feeling that the actions you are undertaking fully influence the sound being created. When an audience is faced with a piece of work with minimal interaction, they begin to further test the limits of the work. This will also lead them to question the authenticity of the composer and the work that they are creating, causing them to lose faith in the work or the artist.

The audience places their trust in the creator of the work that what they are witnessing is a truly interactive piece of music. Therefore, there are ethical implications if the composer misrepresents the level of interaction within the work. Interacting with a piece of work, particularly a public piece of sound work, can be a very vulnerable position to put a person in. They have to be seen to be committing to the work fully and completely to cause the expected interactions. This can include large gestures or interactions with strangers. If this has been misrepresented by the composer to suggest that the work is more interactive than it actually is, then the work is not a fair and equal collaboration. The audience is expected to give themselves to the work whilst being treated as an equal.

In the framings of authenticity for interactive sound and music, the term authentic is being applied with the following expectations in mind:

- The composer intends for interaction to be fully embedded in the work, it is not an add on and it does not mislead the audience about the amount of interaction
- The audience or instigator of the interaction should be central to the work as a collaborator and the work should have been designed with them in mind from the beginning
- The interaction truly impacts the work, having a long-term lasting effect on the piece that can be seen by the audience through their actions

Conclusion

When considering if a piece of work is truly interactive, a number of practical and social considerations need to be taken into account including aspects of musicking and the question of authenticity within the musical work.

Interactivity works on a scale of interaction that moves from attendance, where the audience undertakes traditional audience roles, moving through to participatory and then collaborative roles.

While there may be some practical reasons to keep a work within the attendance or participatory roles, truly interactive work needs the composer to move away from controlling all aspects of the work, moving the audience into the role of collaborators. For the composer this really involves adjusting how they view their role within the work. The shift to more interaction in sound and music, particularly in games, has marked a change in audience expectations and in the hierarchies of the work that we create. This is where the word 'participatory' becomes a challenge for interactive pieces. Participatory art suggests that the composer is allowing the audience into their space, they have created the conditions and the rules for the audience. If a piece of interactive sound and music is truly collaborative then the composer needs the audience to fully realise the work. They are equal in its creation.

The composer has an ethical responsibility towards the audience with how they represent the work. They have entered into a social contract and must hold themselves accountable to their collaborators. Unless there is a justifiable narrative or musical reason to limit the reaction, while presenting an interactive piece of work the composer should ensure that they are fulfilling these obligations to their collaborators.

We can assume, through viewing audience behaviours testing the limits of a work or considering how long they stay, that an audience can tell if a work that has been advertised as interactive is actually limiting the interactive processes. However, this is really beside the point. This assumes that the composer is waiting to be caught out and is hoping to get away with less. We have seen cases of deliberate misrepresentation within work that is not linked to an artistic or narrative reasoning, the Willy Wonka Experience in Scotland being a good example of misleading an audience about how immersive or interactive an experience will be,[38] but these generally represent a more corporate approach to building interactive and immersive experiences. The composer's intention when creating interactive sound and music is to work with the audience to create something new and interesting. Unless working towards a larger artistic goal within the piece, there is no benefit to a composer in misleading an audience about the amount of interaction within their work. Therefore, the goal is not to determine whether the work is interactive, or whether the audience can tell. The goal is to consider how to make the interaction the most meaningful for the benefit of the work and the audience.

Reading group questions

1   What examples are there of work where the interaction does not impact the outcome of the final music? Does this impact your overall enjoyment of the work?
2   When using an interactive work, how does an audience indicate that they do not feel a work is truly interactive?

3 What examples are there when the interaction in a work has been effectively limited? How does this impact an audience's understanding of the work?
4 Does the composer have a social and ethical obligation to the audience or performers in linear, non-interactive work?
5 What is the impact to a composer of an audience losing trust in them?
6 What ethical obligations does the composer have to non-human interactors such as animals or AI collaborators?
7 How does the role of ownership of work change in long-distance collaborations? Can a composer be justified in feeling that they own the work in its entirety (as in the Stockhausen example)? How does this impact the work?
8 How can a composer ensure that an audience knows their actions are impacting the interaction while maintaining the integrity of the work?
9 What are the conflicts, if any, between commercial aims and artistic aims for interactive sound and music?
10 What are the implications of moving from the word 'participatory' to 'collaborative' for a piece of interactive sound and music?

Notes

1 Pierre Boulez, 'Boulez - Anthèmes 2 for Violin and Live Electronics', *Universal Edition* <https://www.universaledition.com/en/Works/Anthemes-2/P0006667> [accessed 19 March 2024].
2 Dan Overholt provides a comprehensive overview of how this can be achieved through interface design. Dan Overholt and Michael Filimowicz, 'Designing Interactive Musical Interfaces', in *Designing Interactions for Muisc and Sound* (Abingdon, Oxon: Focal Press, 2022), pp. 20–47.
3 MiMu, 'MiMU — Music Through Movement' <https://www.mimugloves.com/> [accessed 26 March 2024].
4 Katja Kwastek, *Aesthetics of Interaction in Digital Art* (Cambridge, United States: MIT Press, 2013), p. xvii.
5 Céleste Boursier-Mougenot, *From Here to Ear* (The Curve, Barbican Centre, 2010).
6 Sun Yuan and Peng Yu, *Can't Help Myself* (Guggenheim Museum, 2016).
7 Hatnote, *Hatnote Listen to Wikipedia* <http://listen.hatnote.com/#> [accessed 15 March 2024].
8 Sherry R. Arnstein, 'A Ladder of Citizen Participation', *Journal of the American Institute of Planners*, 35.4 (1969), 216–24.
9 'Radiohead App – Universal Everything – Polyfauna', *Universal Everything* <https://www.universaleverything.com/collaborations/polyfauna> [accessed 19 March 2024].
10 Simogo, *Sayonara Wild Hearts* (Annapurna Interactive, 2019).
11 Thatgamecompany, *Journey* (Sony Interactive Entertainment, 2012).
12 This will be discussed further when discussing the practicalities of composition in chapter 3.
13 Gabe Cuzzillo, *Ape Out* (Devolver Digital, 2019).
14 Reactable, 'Welcome', *Reactable Legacy* <https://reactable.com/> [accessed 26 March 2024].
15 The role of the composer and their intention for the work will be discussed further in chapter 6 within a *werktreue* framing.

16　Arnstein.
17　Claire Bishop, *Artificial Hells: Participatory Art and the Politics of Spectatorship* (Verso, 2012), p. 283.
18　Karlheinz Stockhausen, *Aus Den Sieben Tagen*, 1968.
19　Robin Maconie, *Other Planets: The Complete Works of Karlheinz Stockhausen 1950–2007* (Blue Ridge Summit, United States: Rowman & Littlefield Publishers, Incorporated, 2016), p. 286.
20　Robert P. Morgan, 'Stockhausen's Writings on Music', *The Musical Quarterly*, 75.4 (1991), 194–206 (p. 203).
21　Studio Roosegaarde, *Dune* (Lumiere, Durham, 2009).
22　Random International, *Rain Room* (The Curve, Barbican Centre, 2013).
23　Barbican, 'Rain Room Press Release', 2013 <https://web.archive.org/web/20170325011027/http://www.barbican.org.uk/news/artformnews/art/visual-art-2012-random-internati> [accessed 3 April 2024].
24　Mountains, *Florence* (Annapurna Interactive, 2020).
25　Sun Yuan and Peng Yu, *Can't Help Myself* (Guggenheim Museum, 2016).
26　A pedestrian crossing is actually an interesting case in perceived interactivity as many of the buttons are a placebo for pedestrians and the lights are actually on a timer. The repeated button pressing is actually making no difference to the light sequence, but is seen by the pedestrian as the instigating factor. Tom de Castella, 'Does Pressing the Pedestrian Crossing Button Actually Do Anything?', *BBC News*, 4 September 2013 <https://www.bbc.com/news/magazine-23869955> [accessed 30 March 2024].
27　*The Legend of Zelda: Breath of the Wild* (Nintendo, 2017).
28　Simogo.
29　Steam, 'Sayonara Wild Hearts' <https://store.steampowered.com/app/1122720/Sayonara_Wild_Hearts/> [accessed 3 April 2024].
30　Gabe Cuzzillo.
31　Steam, 'Ape Out' <https://store.steampowered.com/app/447150/APE_OUT/> [accessed 3 April 2024].
32　Yayoi Kusama, *Obliteration Room* (Tate Modern, 2022).
33　Daily tous les jours, *Work* <https://www.dailytouslesjours.com/en/work> [accessed 3 April 2024].
34　Julian Schaap and Pauwke Berkers, '"You're Not Supposed to Be into Rock Music": Authenticity Maneuvering in a White Configuration', *Sociology of Race and Ethnicity*, 6.3 (2020), 416–30 <https://doi.org/10.1177/2332649219899676>.
35　Orchestra of the Age of Enlightenment, 'Our Story' <https://oae.co.uk/who-we-are/our-story/> [accessed 3 April 2024].
36　This will be discussed further in chapter 6 when considering how an artefact can be created from a piece of interactive sound and music.
37　Walter Benjamin, *The Work of Art in the Age of Its Technological Reproducibility, and Other Writings on Media*, ed. by Michael W. Jennings, Brigid Doherty, and Thomas Y. Levin, trans. by Edmund Jephcott and others, 2008 edn (Cambridge, Massachusetts: The Belknap Press of Harvard University Press, 1935), p. 31.
38　Libby Brooks and Libby Brooks Scotland correspondent, 'Glasgow Willy Wonka Experience Called a "Farce" as Tickets Refunded', *The Guardian*, 27 February 2024 <https://www.theguardian.com/uk-news/2024/feb/27/glasgow-willy-wonka-experience-slammed-as-farce-as-tickets-refunded> [accessed 7 April 2024].

References

Arnstein, Sherry R., 'A Ladder of Citizen Participation', *Journal of the American Institute of Planners*, 35. 4 (1969), 216–224.

Barbican, 'Rain Room Press Release', 2013 <https://web.archive.org/web/20170325011027/http://www.barbican.org.uk/news/artformnews/art/visual-art-2012-random-internati> [accessed 3 April 2024].

Benjamin, Walter, *The Work of Art in the Age of Its Technological Reproducibility, and Other Writings on Media*, ed. by Michael W. Jennings, Brigid Doherty and Thomas Y. Levin, trans. by Edmund Jephcott, Rodney Livingstone, Howard Eiland2008 edn (Cambridge, Massachusetts: The Belknap Press of Harvard University Press, 1935).

Bishop, Claire, *Artificial Hells: Participatory Art and the Politics of Spectatorship* (Verso, 2012).

Brooks, Libby, 'Glasgow Willy Wonka Experience Called a "Farce" as Tickets Refunded', *The Guardian*, 27 February2024 <https://www.theguardian.com/uk-news/2024/feb/27/glasgow-willy-wonka-experience-slammed-as-farce-as-tickets-refunded> [accessed 7 April 2024].

de Castella, Tom, 'Does Pressing the Pedestrian Crossing Button Actually Do Anything?', *BBC News*, 4 September2013 <https://www.bbc.com/news/magazine-23869955> [accessed 30 March 2024].

Daily tous les jours, 'Work' <https://www.dailytouslesjours.com/en/work> [accessed 3 April 2024].

Kwastek, Katja, *Aesthetics of Interaction in Digital Art* (Cambridge, United States: MIT Press, 2013).

Maconie, Robin, *Other Planets: The Complete Works of Karlheinz Stockhausen 1950–2007* (Blue Ridge Summit, United States: Rowman & Littlefield Publishers, Incorporated, 2016).

MiMu, 'MiMU — Music Through Movement' <https://www.mimugloves.com/> [accessed 26 March 2024].

Morgan, Robert P., 'Stockhausen's Writings on Music', *The Musical Quarterly*, 75. 4 (1991), 194–206.

Orchestra of the Age of Enlightenment, 'Our Story' <https://oae.co.uk/who-we-are/our-story/> [accessed 3 April 2024].

Overholt, Dan, and Michael Filimowicz, 'Designing Interactive Musical Interfaces', in *Designing Interactions for Muisc and Sound* (Abingdon, Oxon: Focal Press, 2022), pp. 20–47.

Reactable, 'Welcome', Reactable Legacy <https://reactable.com/> [accessed 26 March 2024].

Schaap, Julian, and Pauwke Berkers, '"You're Not Supposed to Be into Rock Music": Authenticity Maneuvering in a White Configuration', *Sociology of Race and Ethnicity*, 6. 3 (2020), 416–430, doi:10.1177/2332649219899676.

Steam, 'Ape Out' <https://store.steampowered.com/app/447150/APE_OUT/> [accessed 3 April 2024].

Steam, 'Sayonara Wild Hearts' <https://store.steampowered.com/app/1122720/Sayonara_Wild_Hearts/> [accessed 3 April 2024].

Media and art examples

Boulez, Pierre, *Boulez - Anthèmes 2 for Violin and Live Electronics*, Universal Edition <https://www.universaledition.com/en/Works/Anthemes-2/P0006667> [accessed 19 March 2024].

Boursier-Mougenot, Céleste, *From Here to Ear* (The Curve, Barbican Centre, 2010).

Cuzzillo, Gabe, *Ape Out* (Devolver Digital, 2019).

Hatnote, *Hatnote Listen to Wikipedia* <http://listen.hatnote.com/#> [accessed 15 March 2024].

Kusama, Yayoi, *Obliteration Room* (Tate Modern, 2022).

Mountains, *Florence* (Annapurna Interactive, 2020).

Random International, *Rain Room* (The Curve, Barbican Centre, 2013).

Simogo, *Sayonara Wild Hearts* (Annapurna Interactive, 2019).

Stockhausen, Karlheinz, Aus Den Sieben Tagen, 1968.

Studio Roosegaarde, *Dune* (Lumiere, Durham, 2009).

Thatgamecompany, *Journey* (Sony Interactive Entertainment, 2012).

The Legend of Zelda: Breath of the Wild (Nintendo, 2017).

Universal Everything, *Radiohead App – Universal Everything – Polyfauna* <https://www.universaleverything.com/collaborations/polyfauna> [accessed 19 March 2024].

Yuan, Sun and Peng Yu, *Can't Help Myself* (Guggenheim Museum, 2016).

Chapter 2

How do audience expectations change in an interactive space?

In interactive music the audience is acting in a role where they are collaborating with the composer through their actions. As previously established in chapter 1, the work cannot exist without an audience and is reliant on the audience fully engaging with an interactive space.

While acting in roles of collaborators, an audience's expectation for an interactive work will change. They will expect to be fully engaged with the work and take on an active and exploratory role, as outlined within the scale of participation.[1] In order to create a piece of work that will fully meet an audience's needs and realise the artistic interactive aims, a composer needs to build an understanding of these audience expectations and behaviours.

This chapter will look at the central role of audience in interactive sound and music with an understanding of audience behaviour and structures of musicking. It will take into account how audiences react to different sonic situations and how they approach an interactive space.[2] A successful composer will use this knowledge in the development of their project to provide layers within their compositional processes allowing different musical audiences to approach the space in different ways.

Audience types and expectations

As interactive sound and music sits within the experience economy,[3] audiences approach a work with a certain amount of expectations. Within our current entertainment landscape audiences have a wide variety of choice, particularly with the amount of home entertainment options available. Using streaming platforms they can access most film and television back catalogues on demand, including films that are playing concurrently in cinemas. Because of this, people need to make an active choice to engage with new culture outside of the home, it is no longer the default option.

According to Audience Answers' Cultural Participation Monitor, across all audience types, people prefer to see live events in person rather than through a digital platform.[4] This data looked at traditional performance styles, such as concerts, ballet etc., but suggests that audiences get enjoyment from experiencing

DOI: 10.4324/9781003344148-3

work alongside other audience members or by being in the space with the 'aura' of the performer, supporting Walter Benjamin's views on the importance of the live person and performer.[5] Younger audiences also pointed to a preference for relaxed social rules within a live environment where it was acceptable to take photos and videos, which contrasted with other audience preferences for 'in the moment' behaviour where they can focus on enjoying the live event as it happens without any phone usage.[6]

Both of these preferences suggest a choice that audiences make when they are engaging with an interactive piece of work. Audiences are looking for an environment that provides them with an alternative to the streaming opportunities that they get at home. This can be in the format of a sharable event, recorded for social media sharing or an event experienced in the moment without a record being kept. While we have seen an increase in digital experiences at home, they have not replaced an audience's need for more human engagement and work that requires an active choice to participate.

Engaging with interactive games at home is also an active choice that people are making. This requires an engagement at home that does not always adhere to the 'two screen' mode of viewing, where a person is watching media and scrolling on their phone at the same time (although a person's second screen when watching a TV show or film may be a mobile game). When engaging with a game, generally the player has to commit fully to only accessing one screen and choosing to immerse themselves fully within the game and its required actions.

In both cases of in person and at home interactive work the audience is aiming for an experience that fully engages them and immerses them in the work. This is supported by the collaborative nature of interactive sound and music, which requires the audience's full attention.

Cultural changes in the roles of adults in society have impacted the audience types for interactive work with immersive theatre and experiences, escape rooms and video games all aiming for an adult audience. As Newzoo reports in their 2024 survey, across generations the majority of the population engage with games. In the Gen Z population (born 1995–2009), 90% of the population engage with games. In the Millenial population (1981–1994), 82% play games, and in Gen X (1965–1980) 67% of the population game.[7] For in person experiences nostalgia and group experiences are key drivers in events, as exemplified by interactive attractions such as The Crystal Maze Experience, which faithfully recreates the 90s TV show team experience.[8]

These trends are significant as this points to the market for experience events and the experience economy. In 2000 Jeffrey Jensen Arnett pointed towards a new life stage of 'emerging adulthood' where young people were leaving education and choosing to delay starting a family, giving them disposable income for experiences.[9] It appears that since this point we have reached an acceptance of adulthood that allows for engagement in the experience economy and events. These are audience members who are fully realised as adults and are choosing to spend their disposable income on interaction, experience and playful activities.

Musicking and participation in an interactive context

Within interactive sound and music the audience holds a central role which links to Christopher Small's concept of musicking, which considers how audiences actively engage and interact with music through listening, community and performance.[10] By allowing us to consider music in a social and community context, Small has created a way for us to place the audience actively within the work as listeners. For example, singing along can be considered musicking, headbanging within the complex time signatures at a mathrock gig can be considered musicking and in a classical concert understanding when to clap (and when not to).

While musicking builds on a sense of community in music, it does still act within some form of hierarchy by placing the artist or performer at the centre of the work and actions of audience members within that community radiating out from that central action. The artist or musician is an instigator for the community. For example, in a classical music orchestral setting it is expected that you will be seated for the performance and (very generally speaking) not clap between movements, while at a pop concert musicking would involve singing along with tracks.

While these are participatory behaviours, they are built on a hierarchy of performer and audience.

This is solidified by the positioning of space with either elevated stages or seating facing a central focus.

In most music performance contexts, the rules of the community are generally controlled by the performer or creator. Dependent on how audience communities are engaged this form of musicking sits between informing and partnership on Arnstein's Ladder of Citizen Participation.[11] For example, if we look at Beyoncé's tour, musicking includes being informed of a dress code for her birthday gig and the viral mute challenge where crowds compete to be the most quiet audience during her performance of the song *Energy*.[12] There are also numerous dance challenges with audiences recreating Beyoncé's dances through social media.[13] All of these examples still place Beyoncé at the centre of the behaviour, it is all in reaction to her music and actions.

The act of musicking in an interactive sound environment changes due to the very nature of the music being created. Arguably, interactive music is a context that cannot exist without an audience making concepts of musicking more prevalent. The work has been created within the context of having an audience (either a gamer or a physical in person audience) at the centre triggering and controlling the work. Without the audience, there is no inciting action for the audio to start or for musical processes to occur. If an interactive work exists without an audience, can the work really exist? Therefore, in an interactive audio context, musicking becomes the cultural engagement with the work and the participatory action required to interact with the work. This aligns with the Cultural Participation Monitor reporting on audience's preferences for live events and relaxed environments, which is looking for a live community aspect to musicking.

In this context the hierarchy of the artist/performer and audience disappears, this shifts the audience's role in the music and the place on Arnstein's ladder of participation. Due to the nature of interactive music it is necessary for the audience to be on the rungs of participation, delegated power and citizen control.[14] This means the artist or composer surrendering a certain amount of control that they have over the work. In terms of roles composers are considered a form of auteur, they are responsible for the overall sound of the work. There are exceptions to this in terms of improvisational practices in jazz or contemporary classical music. However, in an interactive context, this hierarchy shifts and the audience become collaborators in the project. The work in its very essence relies on the audience in the centre. The implications of this for the composer will be discussed in chapter 3, while this chapter focuses on the changes for the audience and their behaviour when this collaborative process is applied.

Space and imagination

As the hierarchy of the composer and audience shifts within an interactive environment, the space within an interactive piece becomes a signifier of the changing roles and expectations for the work. Within the space the audience members move from their everyday roles across to assuming a role of performer and collaborator within the work. This requires almost a sense of make believe when entering the space. The audience must buy into the conceit of the interactive world.

By moving into the space the audience members must let go of the conditions of the world that they have left and accept the interactivity in the space. Accepting the world as created by the composer and creating a sense of agency for their actions. This requires an imaginative leap of not quite immersion, but a space where they can see the reality of their world and accept the conditions of the space.

The audience is required to be active in this use of the imagination. While the composer can set the conditions for this make believe to take place, they cannot ensure that it truly happens unless the audience commits to the work, that is an act of faith that they take in the audience as their collaborators.

In *An Actor Prepares* Stanislavski, speaking as the Director, outlines the difference between imagination and fantasy to an artist:

> Imagination creates things that can be or can happen, whereas fantasy invents things that are not in existence, which never have been or will be. And yet, who knows, perhaps they will come to be. When fantasy created the Flying Carpet, who could have thought that one day we should be winging our way through space? Both fantasy and imagination are indispensable to a painter.[15]

In our interactive space the composer uses both fantasy and imagination by creating the world in which the work will take place, building a full picture of the work and its boundaries, the conditions for the interaction. The audience then takes on the role of the actor, filling in the inner logic of the scene and the interactive work in how they engage with the piece.

The composer is responsible for creating tangible conditions for this audience's imagination within the interactive space. Unlike the actor in Stanislavski's example, the audience may not have the opportunity to prepare for their interaction and will likely be encountering the conditions for interaction the first time that they see the space.

Stanislavski outlines the conditions where imagination can fail:

> In the first place, you forced your imagination, instead of coaxing it. Then, you tried to think without having any interesting subject. Your third mistake was that your thoughts were passive. Activity in imagination is of utmost importance. First comes internal, and afterwards external action.[16]

The audience looks to the composer to create some of these conditions to support their interaction with a work. The space should be conducive to imagining the conditions of the work giving clues to the understanding of space, character and the expectations of them as a collaborator. The space should allow for active imagination and for them to build their internal understanding before translating that to external interaction with the piece.[17]

The entry space can act as a boundary in which the conditions for imagination take place, acting as a liminal place between the real world and fantasy. How an audience is prepared for a move to an interactive space will set the conditions for their understanding of the world and the interactions that are expected of them. This can be seen very clearly in theme park queues, where atmospheric music is used to set the tone for the ride. Combined with being able to see the ride ahead, the queue acts a liminal space before the ride.

Audience expectations and approaches to space

Audiences for interactive sound and music arrive at an interactive space with an expectation for individualised experiences. A number of installations and interactive events are built with a group or community level of participation in mind, including large-scale immersive theatre events and Massively Multiplayer Online Games (MMOs). While a person may be participating as part of a group, the audience expectations in these spaces is that they will have an individual and personalised experience and in these spaces the whole structure of the experience is built around personal moments. For example, Punchdrunk audience members can randomly be selected for a one-to-one interaction with an actor in a small room. Audience members then begin to seek out these experiences; for example, on the way out of Punchdrunk's *The Burnt City*[18] I heard a conversation between two audience members where one had been invited to an individual experience with a character and their friend was wondering what he could have done differently so that he could also have had an individual experience, commenting that he'd 'tried everything' to let the characters know he wanted to interact with them individually.

This is described by Keren Zaiontz as 'narcissistic spectatorship', which

> ...encourages the viewer to fully engross herself in an artistic production in a way that highlights her own singular relationship to the piece. The spectator is not positioned as an author or agent who has the power to create or enact concrete change, but as an *experiencer* of the piece.[19]

This focus on individualisation of experience is also a key marketing tool for immersive and interactive events. Many immersive and interactive productions are built and marketed on the concept of FOMO – Fear Of Missing Out. Audience members will tweet and share experiences but no photos can be taken during the productions, e.g. the Secret Cinema terms and conditions explicitly state that 'visitors' phones, cameras, and other recording equipment will be locked in a non-transparent bag for the duration of the Event and must not be used'.[20] This preserves other audience members' experiences of an event, making sure that they are not distracted by phones and cameras but also acts as a marketing tool. If people want the experience they must attend the event. In addition, audience members after may compare the elements that they experienced within a production based on the route that they took. For example, in Secret Cinema productions there are regularly sub events that happen within the productions (run in the similar way to a side quest) that can only be accessed if audience members have access to key words, tickets or are standing in the right place at the right time. Much like with uncovering rare trophies, these small and individualised experiences become key to the production, as Gareth White described an encounter at Punchdrunk's 2009 production *Tunnel 228*: 'I was pleased with the way I played this game – with some of the lines I invented for myself, as well as my evasiveness – and was also delighted to have an encounter all of my own in a space not seen by most of the audience'.[21]

Equally, when people approach a space with individualised experiences in mind but also within a social group this impacts the interaction design and types of interaction within a space. There are many sound installations that are built with collaborative aims in mind where strangers within a space are expected to work together to produce musical results, for example by coordinating their movement within the space, joining hands with strangers etc. In reality, not everyone enters a space looking for a social interaction with strangers, meaning that people are being pushed beyond their comfort zone of interaction. When you observe these styles of interaction in action, you observe groups of people who have joined a space together, e.g. families or friendship groups, that will only participate and collaborate within those friendship groups. There is a reluctance to work in a wider group of strangers. This leads to whole parts of the interaction not being realised or becoming fragmented. There are some sound installations that have successfully addressed this through the interaction design by designing individual interactions that combine to create the piece as a whole, allowing for individualised or social group experiences while adding to the overall effect of the work. For example, *21 Swings* by Daily tous les jours[22] provides a strong example of how interaction can be based on smaller interactive experiences that build to a large collaborative piece. The installation consists of 21 swings that sit alongside a road and how audience members play on the swings impacts on the overall sound. If the people on the swings choose to synchronise their actions they unlock a range of musical patterns, while if they choose to work against other people on the swings they get a different pattern. Similarly, Umbrellium's *Assemblance*[23] is made up of multiple individual sound and light experiences. Each person can interact, moving the sound and light as an individual and building the overall experience without having to join with groups of strangers, unless they wish to.

Encouraging interaction

Another consideration when looking at how audiences approach the space is the permission to interact with the space. What I've found through working within interactive sound installations is that audiences are sometimes reticent to approach a space and, adults in particular, don't enjoy being told just to go and interact. There is a difference between how adults and children approach an interactive space. Years ago, I created an interactive installation for a youth organisation that had adult volunteers. The installation was built from a through composed piece of electronic music and used an Xbox Kinect to provide a live, adaptive mix of the work through audience movement. The intention was that the interaction would be accidental and triggered as audience members walked through the space. While children were happy to cut through the space and test the sound to see what would happen, I observed that adults would hold back the children from interacting, telling them not to enter the space until they had received the full instructions for interaction. The adults didn't like that the space was built around free exploration and they needed more guidance of how to behave in a space. This was rectified in a very simple, very quick way by putting sticky tape arrows on the floor but it did get me thinking about the very interesting way that adults approach space, especially if they're not used to free exploration and need to be told the right way to interact with the work.

Chapter 1 outlined the social contract and ethical obligations that the composer has in place with the audience as collaborators when creating a piece of interactive sound and music. These obligations extend to the environment that the audience is placed in for the interaction. Engaging with an interactive work can be a vulnerable position for an audience member who is not used to working in an interactive space. Interacting publicly is similar to being expected to perform publicly. It expects the audience member to be visible in the space, interacting using exaggerated gestures or committing to a fictional narrative. Bishop refers to this in terms of 'delegated performance'[24] where the audience are delegated to by the artist to undertake their role within the work. In an interactive sound installation, the audience is taking the roles of both the performers and the composer in completing the work as collaborators and realising the role as performers. They are also at the centre of the work and are being viewed by other audience members in the same way that they would view a professional performer. Within her discussion of delegated performance, Bishop describes the aim of a 'shared reality between viewers and performers';[25] within Bishop's examples the delegated performer is a non-professional performer or a person from a different discipline who is taking the place of a professional performer. In interactive sound and music, the link between viewers and performers are even closer as they overlap. The audience will be in the role of both performer and viewer simultaneously, engaging with the work and viewing others engaging with the work.

When considering the role of the audience as simultaneously delegated performer and audience, it is often the role of the composer to create a safe environment for this interaction to take place that supports the audience with how to engage without feeling a risk of judgement. A safe environment is an ethical responsibility that the composer holds within their work and suggests not just a physical safety complying with legislation but also psychological safety. This ethical requirement for psychological safety may not be agreed by all artists and there can be places where an artist plays on this concept of the safe environment for interaction, creating a work that is perceived as being unsafe. For example, Marina Abramović and Ulay's *Imponderabilia*[26] consists of two performers of different genders who stand in a doorway facing each other. In order to access the main exhibition the audience have to pass through the doorway and make a choice to be face to face with one of the performers. Amabramović described her intention to ensure that the audience is forced to make this choice by narrowing the doorway so that they could not walk forward and avoid the choice.[27] However, in the career retrospective held at the Royal Academy of Arts in London in 2023, different decisions were made in order to ensure audience comfort with the project, these included providing a separate entrance for people who choose not to engage with Imponderabilia and contact information for people who wanted to discuss the content before attending the exhibition.[28] By providing the audience with a choice they still engaged with the concept of the work, providing an ethical environment while still including artistic challenge.

Anonymity can be an effective tool to support an audience to interact in a space. It removes the risk of judgement that a person may fear when wholeheartedly committing to interacting within a space, as demonstrated through Punchdrunk's use of masks which White points out disinhibits the audience and allows them to focus on the immersive event.[29] However, this anonymity requires a very structured approach to the social contract provided within the work by the creator and the provision of equipment for the entire audience.[30] Masks act as a costume within the interactive work, allowing the audience to assume a character within the piece. Similarly, the costume requirements of companies such as Secret Cinema provide a different option for the audience to assume a character within a space, again signing up to a social contract that they will fully commit to an interactive environment.

To provide a safe environment for an audience to engage with a piece of interactive sound and music, without requiring additional equipment or costuming such as masks, requires an understanding of the concepts of play and playfulness. Audiences may be required to work outside of their comfort zones and push their own approaches to the work; this means they have to begin experimenting in the way that they did as children when exploring and testing. Roger Callois defined play as fulfilling the following conditions:

1   Free: play should not be obligatory and should be entered into freely by those engaging
2   Separate: there is a clear beginning and end to play, and it is fixed within a space
3   Uncertain: the end results should not be determined at the beginning of the play
4   Unproductive: players should end play in the same condition in which they started. Play should not be used for profit making activities but can result in property being won or swapped
5   Governed by rules: rules within play are defined for the duration of the play only. Rules can be redefined during play, for example when children are playing a chase game they can declare an area as 'base' in the middle of the game and, if agreed by all, this rule is then a modification to the game
6   Make-believe: there should be some awareness of a second space or mode during play which sits separate but alongside real life[31]

Within interactive sound and music some concepts of play are applicable. Interaction with the work should be free and open, we expect that there is a clearly defined start and end to the work and the work will take place in a set space and that there should be a sense of defined rules and expectations for how the interaction will work. However not all interactive sound and music is necessarily built around play, particularly more political works as these may deliberately require the audience to stay aware of the world around them, rather than working with make-believe and there may be an expectation of productiveness particularly if there is a political call to action within a work.

What all interactive sound and music does require from the audience is, if not play, a sense of playfulness, through a willingness to engage with the work and experiment with the potential outcomes. Playfulness in this context requires the ability to enter a space and engage with the work with an exploratory mindset allowing the work to unfold through interactive means. This means audiences have to be willing to follow the rules of interaction as set out by the composer, including free exploration if that is what the work determines. They are allowed to subvert these rules and test the limits of the work as an extension of the exploratory playfulness. There is a sense of performativity in this playfulness and exploration in interactive sound and music, as Kwastek raises in her writing on play in relation to play in interactive art.[32] This performativity relates to the playfulness required from an audience within a performance, where the audiences are constructing a reality with their actions. This is taking on a role that would normally be occupied by a performance artist. For some audiences, this performativity may be a challenge as it requires them to take a more central, visible role than they may be used to where they feel their actions may be scrutinised or they may be caught interacting 'incorrectly' with the work.

The question is then how can we ensure that audiences feel that they have the right environment to encourage the sense of playfulness and a safe environment for performativity in an interactive sound installation? Particularly, ensuring that they have the required instructions needed to ensure that they are not worried about interacting 'incorrectly' with the work. This requires supporting all audiences to display the characteristics of gamers in interactive spaces.

Turning theatre and classical audiences into gamers

An immersive, interactive space, outside of a games environment, has a lot formally in common with an open world game. BOP Consulting noted in their report on the state of UK theatre that immersive theatre 'requires new skills and knowledge, which often reside outside the theatre sector (e.g. within performance art or video games)'.[33] While this was outlined with the skillset in mind to create immersive theatre, this has a wider application to audiences and their roles within an interactive space, which requires the composer to have an understanding of performance art and video games. This is due to the narrative design of exploratory immersive theatre and interactive sound installations. In productions with free exploration audience members are given quests or challenges e.g audiences can be tasked with piecing together a fictionalised stories, with each room holding clues and information in the form of letters and artefacts. This is similar in form to games such as *Return of the Obra Dinn*,[34] where you play as an insurance agent searching an empty ship for clues about its history, while productions with combinations of vignettes and free interaction, where scripted scenes are interspersed with interaction, are similar in form to a more traditional open world game where there is an overarching story line to be followed, supported by free exploration, objectives and cut scenes. In addition, interactive installations in their forms are based on individualised experiences. While you may be participating as part of a group, similar to an online open world co-op, the whole structure of the theatre experience is built around individualised moments, for example being rewarded as an audience member with a one-to-one interaction with an actor in a small room. This is similar to online open world games where players are rewarded for individual achievement as well as group achievements.

Adults who regularly play videos games are already practiced in bringing playfulness into their practice. A game audience works with a different set of behavioural norms that assumes more control and ownership of a space (albeit virtual), meaning that they already operate within citizen control on Arnstein's ladder. Within this space the player is expected to take an exploratory role where they smash every pot, check in every chest and go behind every waterfall. Through this they are rewarded by additional interaction, narrative information or collectable items.

Audiences who are not used to playing games will struggle with this expectation of playfulness as they are used to musicking within more traditional audience roles and being given instructions outlining the required participations as outlined in chapter 1. Conditions need to be created to create a safe environment for playfulness. In an interactive space, virtual or physical, we are looking to create a space more similar to a game environment where audiences experience music through more exploratory behaviour. In order to fully experience all that the work has to offer, the audience obviously needs to be willing to interact. When working with an audience of traditional music consumers used to a hierarchy of performer and audience, this becomes particularly challenging as it is going against previously agreed conditions of musicking. For interactive work to be successful audiences have to be encouraged to interact, which means that the design has to influence and encourage this behaviour.

To encourage this behaviour the audience needs to become game literate through the use of familiar game mechanics and approaches that support interaction within a game. This will be discussed in depth from the composer's perspective in chapter 3.

Player motivations

If we accept that the aim is for audiences in all interactive spaces to take similar approaches to game audiences then we must then expect that composers will consider the different approaches and expectations that audiences will bring to an interactive space. If you observe an audience within an interactive narrative space, e.g. an immersive theatre production, you can see how people bring a different approach to every show. Some audience members rush to look through every prop available, trying to piece together clues to the space, others follow particular actors to get a complete storyline, some look to explore every room while some wait for instruction. When in a space with freer interaction, and less reliance on narrative, some audiences will test different objects at random until they encounter interaction, some will look to test the complete interactive limits of a space and others will observe other audience members until they fully understand what is expected within a space.

The different approaches to engaging with interactive space are best explained by applying Quantic Foundry's Gamer Motivations,[35] which outline how a player interacts with a game space and their expectations when approaching a space, to wider interactive sound and music environments. Quantic Foundry's gamer motivations are a market research tool that look at why players enjoy games and how they approach a space. A player can sit across multiple categories of motivations. I would argue that, due to the formal similarities between immersive and interactive works and games, these motivations are equally applicable to other interactive settings:

Action: these players and audiences look for fast-paced games and want to be central to that action e.g. a first-person shooter is an example of gameplay that would appeal to somebody motivated by action. They are central to the space and the plot cannot advance without them. In immersive theatre examples, these audience members actively seek out the plot and follow actors around the space.

Social: audiences enjoy team and social games. This may include games with a chat function or sofa co-op games, games where players are in the same space and play co-operatively on one screen. Outside of games, audience members who have social motivations may either attend with a group or look to build connections with others within a space. In general, in a public art capacity, audience members will look to interact with the group that they have attended with. Social audiences are particularly common in live interactive events where attendance tends to occur within groups.

Mastery: players will look to build skills or solve puzzles. Immersive audiences with this motivation will be found at escape room style experiences where there is an active, puzzle-solving component to the work.

Achievement: these players approach a game looking to complete it as fully as possible including finding all collectibles and earning all trophies. In an immersive theatre context they will be drawn to puzzle-type environments such as escape rooms.

Immersion: players are motivated by complex characters and storylines. They will be drawn to immersive theatre productions that feature large-scale plots and complex sets (e.g. Punchdrunk-style productions).

Creativity: creative players will look to approach and interact with the game around the question 'what if...'; this type of player will appreciate character customization and take an explorative approach to challenges in the game. Audience members driven by creativity may choose a more hands-on approach to an installation or take a different route around a space.

Often audience and player actions within a space can seem unpredictable to a composer, particularly when building interactive sound and music. While a composer can design an interactive piece of work based on their own approaches to a space, they cannot always understand how a stranger would approach a work. If an audience member appears to be breaking a piece of work, by testing all the interactive limits or by moving through the piece at speed, they are not wrong in their approach, they are just approaching the work with a different perspective. The value that they are gaining from the piece is different to that which the composer intended, but it is still a valuable experience. Within a truly collaborative interactive space, the composer has to accept this value to different approaches and design with them in mind.

A challenge for designers of interactive sound and music is understanding the difference in these motivations in a real-life space to build a range of interaction options. A successful work will address multiple player motivations, encouraging people to engage on their own terms. For example, *Untitled Goose Game*,[36] a puzzle game where you play as a goose with a to do list to terrorise a neighbourhood, has puzzles for people motivated by mastery, additional collectable challenges aside from the narrative for people motivated by achievement and hidden experiences and the option to treat the work as a sandbox for people driven by creativity. Additionally, they added a co-op mode for people who are motivated by social engagement.

In interactive audio everything is intentional

With player motivations in mind and the aim to create an environment where the audience are able to undertake a playful, exploratory role, it is important to understand how an audience is listening within a space. In an interactive environment this listening is a critical action; it is a reasonable assumption that the audience are reading into any of the sound within the space to guide their interactions and how they explore a space. Every sound, therefore, must have a practical function built with audience experience in mind.[37]

Based on player motivations, audiences are approaching spaces listening to the sound to support how they wish to engage with the space, with different listening styles in mind as they look for clues to how to interact. This gives sound a very different, practical function. Chion's modes of listening, expanded by Huron and Collins, can provide a framework to understand how audiences interpret sound within an interactive space.[38] Modes of listening suggest that people listen in different ways based on different contexts. These approaches to listening can overlap and sometimes provide multiple functions to the listener. Chion initially wrote about three modes of listening within an audiovisual context, while Huron discussed an additional 21 modes of listening that can be applied across audiovisual media, performance and listening contexts.

Of these modes of listening the following are the most relevant for how an audience member approaches audio in an interactive space:

Causal Listening: Listening to the source of a sound. This can help audience members to move around a space and find clues within a space. For example, if a car alarm goes off on a street, causal listening is used to listen for which car is causing the sound. In an interactive environment, strategically placed sound can be used to guide an audience around the space.

Semantic Listening: Listening for clues in the environment. For example, if morse code is included within a soundtrack, an audience member or player using semantic listening will determine whether a hidden message is being embedded within the music.

Signal Listening: Listening in anticipation and using audio as a guide in a space. Signal listening can involve listening to music or audio clues in order to time actions. For example, rhythm games, where the player must time their actions to the pulse of the music.

When an audience member walks into a space they are using a combination of these modes of listening to understand the space and how they should be interacting with an interactive work. For example, they use causal listening to listen for sources of sound which tell them how to move around the space, they use semantic listening to build an understanding of the location of the work and how they are expected to interact within a space and they use signal listening to work out their expected actions. These elements all work together as part of the audience experience. Composers and sound designers can work with these modes of listening to create an environment that creates a safe environment for interaction. Through this the sound will act as interface for the audience within the space.[39]

Practically speaking, modes of listening mean that in interactive audio, whether the composer and sound designer had intended it that way or not, the audience will read into every sound as being intentional to the space. If sound is placed behind a door, they will consider it a signal that there is an interaction happening behind the door and attempt to go through the door. If there is a repeated pattern within a soundtrack they make take this as a clue for the narrative. This is an area where the composer needs to be careful in their approach to building sound within an interactive space. Misplaced sound, or additions to the work that are intended to sound good but don't have a practical function, can change an audience's view of the work and movement within a space.

The power of a group

Within an interactive space, if a group reaches a certain size the behaviours change. While expecting individualised experiences the audience begins to act as a collective. This has an impact for a composer's understanding of interaction design and how audiences will respond to sound within a space.

Hughes notes that crowds of pedestrians can act like a liquid. Each person holds different characteristics in how they walk and navigate a space but they demonstrate the 'flow of a liquid governed by the frictionless, shallow water equations'.[40] The interesting part of this behaviour is how this liquid-style behaviour adjusts to accommodate different walking speeds and approaches, with faster walkers looking for paths around the flow to keep moving towards their destination. This can be demonstrated on a busy shopping street; while the centre of the pathway is dominated by people walking at a similar pace in the centre, faster walkers will weave in and out of the crowd cutting into the main throng and stepping out into roads. For an interactive space, this has an interesting implication for audience flow. The audience will retain their need for individual interaction and experiences, but will also recognise the flow of the group. The composer needs to avoid bottlenecks within a space, providing different routes around the work so that people can explore at different speeds while maintaining the natural fluidity of the group dynamic. There needs to be space to overtake and cut back into the work without losing the narrative flow and cohesion within the work.

Similarly, a crowd has the capacity to swarm within a space. This is when a crowd works together to demonstrate a similar dynamic to systems demonstrated by insects or birds. Bellomo and Bellouquid provide a number of conditions for modelling swarm behaviour including:

- A defined objective, that may vary during 'panic mode'
- A non-linear common strategy to which all individuals contribute
- Group dynamics that adjust based on where organisms are within a swarm[41]

This swarm behaviour can be seen in a public art or immersive theatre context, through how crowds react to points of interest or action within the event e.g. if there is a timed action or scene being performed that can only be experienced if caught in the moment. This can particularly be seen in events such as immersive theatre where free exploration is interspersed with vignette performances that highlight elements of the narrative. In these cases, audiences can be seen 'chasing' the scene. If one audience member detects that action is about to start and moves with purpose towards a potential scene, they are likely to be followed by other audience members in a swarm-like action. Similarly, audiences can follow individual actors around a space in a swarm in an attempt to locate action. This swarming exemplifies Kaltsa et al.'s model of how swarm behaviour can be used as a tool to detect 'interesting events' within a public space.[42] This also links to concepts of narcissistic spectatorship and the expectation for individualised experiences within an interactive piece of work as audience members 'chase' experiences en masse, ensuring that they are having parity of experience with their fellow audience members.

This swarming behaviour, while interesting to observe within an audience, can provide a challenge within interaction design, particularly when considering larger audiences and the safety of movement within a space. If an audience is all swarming to the same experience this can create a cluster of activity centred around a small area within the space. Audiences may struggle to experience the work or have parity of experience, particularly if they are on the outskirts of the swarm, leading to them feeling that they have not had the same performance of the work as the other audience members. Similarly, if a work is built on a number of consequent vignettes or experiences, the audience may swarm repeatedly around the space causing bottlenecks. This behaviour can be reduced or mitigated through offering repeated and simultaneous opportunities within the work, allowing audience groups to cluster in different spaces within the work.

Crowds can also provide contagious behaviour to a group environment as can be seen in sports crowds. For example, in a stadium, if a team is winning, the crowd will collectively be buoyed by this, while they can also turn on a team if they feel they are not performing. Hill et al. discuss this as a ritualised behaviour within four stages:

- Preparation – where a person prepares for an event and learns the requirements for participation within a private environment or during an event. This can include building expectations prior to an event
- Activating the atmosphere – this starts in smaller groups before attending an event where the atmosphere starts to build, for example when walking towards an event or entering an installation
- Climax – where the atmosphere passes from smaller groups to a large group dynamic
- Recovery – where memories and experiences from an event are stored for use in future events[43]

This contagious behaviour within a crowd is a useful tool for interactive sound installations, particularly when there are audience members who are unused to interacting in a space. If the audience reaches a critical mass of excitement within the interaction, other audience members will follow suit. Similar to the ritualised chants in a football match, they will be able to model the behaviour expected in the space by following the actions of other audience members. This is known as the Honeypot Effect where audiences are drawn to interaction through modelling the behaviours of others and being drawn to an interactive space by the atmosphere generated by other audience members.[44] Additionally, when in a large enough group, interactive behaviours are less exposing and allow self-conscious audience members to 'hide' within the group interaction giving anonymity within the size of the group.

By understanding the larger group dynamics, a composer is able to design with the group behaviours in mind, building in interactions that provide a sense of audience flow through a space with multiple spaces for audience interaction both individualised and within the larger group.

Conclusion

There is no unified audience experience within a piece of interactive sound and music, as a key selling point of interactive work is the individualised experience which audiences seek out in interactive environments, placing themselves at the centre of the work. There are, however, consistencies in how different audiences approach an interactive piece of work.

Within an interactive piece of work the space acts a signifier for audiences to use their imagination and engage with a sense of playfulness, putting themselves in an exploratory mindset. Within this exploration, different motivations will impact how a person engages with a piece of work, including how they test the space and the speed with which they move through an interactive piece of work. Part of this engagement involves listening for clues for how to engage with the interactive space and narrative through approaches such as modes of listening, meaning that within an interactive space all audio is considered intentional by the audience. If this isn't considered within the composer's approach to interactive audio they can mislead an audience with how they move around a space, or cause bottlenecks in audience movement.

As physical interactive environments borrow techniques from games, a game literate audience fares well in physical interactive environments; they are used to using exploratory approaches and testing the limits of environments. While the majority of adults do engage with games, interaction in a public form can be an intimidating environment especially as this places the audience in the role of both the audience and performer, where they are also aware of other people performing around them. Working in larger groups can allow audiences to model behaviours on those around them, demonstrating and encouraging interaction in a work. While counterintuitive, a larger audience provides more anonymity for interaction, supporting audiences to engage further in the work. These larger audiences also have the added benefit of creating a contagious atmosphere within the work, again supporting interactive processes.

Understanding audience behaviours within an interactive space can support the composer when creating their work to consider the impact of their compositional decisions and interactive interfaces. There are ethical implications to creating an interactive environment which include ensuring that audiences are supported in an interactive space, this requires the composer to build their work from an empathetic framework starting from understanding audience behaviours and using this to structure the sound and music and build meaningful, interactive experiences.[45]

Reading group questions

1 Using an example from a work you have seen, or a game you have played, how have you approached a space and what were your aims for the interactive work?

2 With greater access to interactive media, how have audiences expectations changed for interactive work in recent years and how does this impact audiences understanding of space and interaction?

3 How can a liminal space further set expectations for interaction?

4 How can audiences be supported to build imagination and play into their approach to engaging with interactive sound and music?

5 When engaging with a new interactive space, how can an audience understand their role as collaborator?

6 How could a composer working in an interactive context support groups of strangers to engage with each other in a work, that respected their comfort zone of interaction?

7 How do we use different modes of listening when approaching everyday life and how do these change in an interactive space?

8 How can a composer design a space to encourage audience flow, considering larger group dynamics?

9 How can audiences seeking individualised experiences be balanced with the group dynamics through crowd flow, swarm and the honey pot effect?

10 How does the audience understand the musical mechanics and structure within a piece of interactive sound and music?

Notes

1 See chapter 1.
2 This does contravene some structures of composer as the ultimate creator and auteur.
3 B. Joseph Pine II and James H. Gilmore, *The Experience Economy* (Brighton, Massachusetts: Harvard Business Review Press, 2019).
4 Audience Answers, 'Audiences Prefer Live Events to Digital | Audience Answers' <https://evidence.audienceanswers.org/en/evidence/articles/audiences-prefer-live-events-compared-to-digital> [accessed 8 April 2024].
5 Walter Benjamin, *The Work of Art in the Age of Its Technological Reproducibility, and Other Writings on Media*, ed. by Michael W. Jennings, Brigid Doherty, and Thomas Y. Levin, trans. by Edmund Jephcott and others, 2008 edn (Cambridge, Massachusetts: The Belknap Press of Harvard University Press, 1935), p. 31.
6 Audience Answers, 'Younger Audiences Prefer More Relaxed Behavioural Codes | Audience Answers' <https://evidence.audienceanswers.org/en/evidence/articles/younger-audiences-prefer-more-relaxed-behavioural-codes> [accessed 8 April 2024].
7 Newzoo, *How Consumers Engage with Video Games Today: Newzoo's Global Gamer Study 2023*, 2023, p. 9 <https://newzoo.com/resources/trend-reports/global-gamer-study-free-report-2023> [accessed 8 April 2024].
8 Little Lion Manchester, 'The Crystal Maze LIVE Experience in London & Manchester' <https://the-crystal-maze.com/> [accessed 8 April 2024].
9 Jeffrey Jensen Arnett, 'Emerging Adulthood: A Theory of Development from the Late Teens through the Twenties', *American Psychologist*, 55.5 (2000), 469–80.
10 Christopher Small, *Musicking: The Meanings of Performance and Listening* (Wesleyan University Press, 1998).
11 This is outlined further in chapter 1. Sherry R. Arnstein, 'A Ladder of Citizen Participation', *Journal of the American Institute of Planners*, 35.4 (1969), 216–24.
12 *BEYONCÉ MUTE CHALLENGE BATTLE - WHO WON? ATLANTA OR NEW ORLEANS*, dir. by Foodie in Lagos, 2023 <https://www.youtube.com/watch?v=b6ZayFYSHcs> [accessed 18 March 2024].
13 *Beyonce CUFF IT Dance Challenge #TikTokHypeComps*, dir. by TikTok Hype Compilations, 2022 <https://www.youtube.com/watch?v=v-GMPNe-qRE> [accessed 18 March 2024].
14 Personally, I am wary of the term participation in interactive work as this still requires mediating through a central figure of power.
15 Konstantin Stanislavski, *An Actor Prepares*, trans. by Elizabeth Reynolds Hapgood, 2008 edn (London: Methuen Drama, 1937), p. 60.
16 Stanislavski, p. 63.
17 These conditions for meaningful interaction will be discussed further in chapter 5.

18 Punchdrunk, *The Burnt City* (Woolwich, 2022).
19 Keren Zaiontz, 'Narcissistic Spectatorship in Immersive and One-on-One Performance.', *Theatre Journal*, 66.3 (2014), 405–25 (p. 407).
20 Secret Cinema, 'Secret Cinema - Terms & Conditions' <https://www.secretcinema.com/terms-conditions> [accessed 29 March 2024].
21 Gareth White, 'On Immersive Theatre', *Theatre Research International*, 37.3 (2012), 221–35 (p. 231).
22 Daily tous les jours, '21 Balançoires (21 Swings) | Daily Tous Les Jours' (Montreal, 2011) <https://www.dailytouslesjours.com/en/work/21-swings> [accessed 15 March 2024].
23 Umbrellium, *Assemblance* <https://umbrellium.co.uk/projects/assemblance/> [accessed 29 March 2024].
24 Claire Bishop, *Artificial Hells: Participatory Art and the Politics of Spectatorship* (Verso, 2012), chap. 8.
25 Bishop, p. 239.
26 Marina Abramovic and Ulay, *Imponderabilia*, 1977.
27 The Museum of Modern Art, 'Marina Abramović and Ulay. Imponderabilia. 1977/2010 | MoMA' <https://www.moma.org/audio/playlist/243/3119> [accessed 7 April 2024].
28 Royal Academy of Arts, 'Marina Abramović | Exhibition | Royal Academy of Arts' <https://www.royalacademy.org.uk/exhibition/marina-abramovic> [accessed 7 April 2024].
29 Gareth White, 'On Immersive Theatre', *Theatre Research International*, 37.3 (2012), 221–35 (p. 224).
30 And very clear disinfection processes for any shared equipment.
31 Roger Caillois, *Man, Play and Games*, ed. by Meyer Barash, 2001 edn (Urbana and Chicago: University of Illinois Press, 1961), pp. 9–10.
32 Katja Kwastek, *Aesthetics of Interaction in Digital Art* (Cambridge, United States: MIT Press, 2013), pp. 84–85.
33 B.O.P. Consulting and Brian Devline Associates, *Arts Council England: Analysis of Theatre in England*, 2016, p. 5.
34 Lucas Pope, *Return of the Obra Dinn* (3909, 2018).
35 Nick Yee, 'Gaming Motivations Group Into 3 High-Level Clusters', *Quantic Foundry*, 2015 <https://quanticfoundry.com/2015/12/21/map-of-gaming-motivations/> [accessed 12 March 2024].
36 House House, *Untitled Goose Game* (Panic, 2019).
37 Michiel Kamp's framework for listening using 'semiotic hearing' is also a useful framework for understanding how audiences approach an interactive space. Michiel Kamp, *Four Ways of Hearing Video Game Music* (Cambridge University, 2015).
38 Michel Chion, *Audio-Vision: Sound on Screen*, 2nd edn (Columbia University Press, 1990), pp. 25–34; D. Huron, 'Listening Styles and Listening Strategies', in *The Society for Music Theory*, 2002; Karen Collins, *Playing with Sound: A Theory of Interacting with Sound and Music in Video Games* (MIT Press, 2013), pp. 4–15.
39 This will be outlined further in chapter 4.
40 Roger L. Hughes, 'The Flow of Human Crowds', *Annual Review of Fluid Mechanics*, 35 (2003), 169–82 (p. 173).
41 Nicola Bellomo and Abdelghani Bellouquid, 'On the Modeling of Crowd Dynamics: Looking at the Beautiful Shapes of Swarms', *Networks and Heterogeneous Media*, 6.3 (2011), 383–99 (p. 396) <https://doi.org/10.3934/nhm.2011.6.383>.
42 Vagia Kaltsa and others, 'Swarm Intelligence for Detecting Interesting Events in Crowded Environments', *IEEE Transactions on Image Processing*, 24.7 (2015), 2153–66 <https://doi.org/10.1109/TIP.2015.2409559>.
43 Tim Hill, Robin Canniford, and Giana M. Eckhardt, 'The Roar of the Crowd: How Interaction Ritual Chains Create Social Atmospheres', *Journal of Marketing*, 86.3 (2022), 121–39 (pp. 125–31) <https://doi.org/10.1177/00222429211023355>.

44 Niels Wouters and others, 'Uncovering the Honeypot Effect', delivered at 'Proceedings of the 2016 ACM Conference on Designing Interactive Systems - DIS '16', 2016 <https://doi.org/10.1145/2901790.2901796>.
45 These will be discussed further in chapters 4 and 5, using this understanding of audience behaviour as a basis for practical considerations.

References

Arnett, Jeffrey Jensen, 'Emerging Adulthood: A Theory of Development from the Late Teens through the Twenties', *American Psychologist*, 55.5 (2000), 469–480.

Arnstein, Sherry R., 'A Ladder of Citizen Participation', *Journal of the American Institute of Planners*, 35. 4 (1969), 216–224.

Audience Answers, 'Audiences Prefer Live Events to Digital | Audience Answers' <https://evidence.audienceanswers.org/en/evidence/articles/audiences-prefer-live-events-compared-to-digital> [accessed 8 April 2024].

Audience Answers, 'Younger Audiences Prefer More Relaxed Behavioural Codes | Audience Answers' <https://evidence.audienceanswers.org/en/evidence/articles/younger-audiences-prefer-more-relaxed-behavioural-codes> [accessed 8 April 2024].

Bellomo, Nicola, and Abdelghani Bellouquid, 'On the Modeling of Crowd Dynamics: Looking at the Beautiful Shapes of Swarms', *Networks and Heterogeneous Media*, 6. 3 (2011), 383–399, doi:10.3934/nhm.2011.6.383.

Benjamin, Walter, *The Work of Art in the Age of Its Technological Reproducibility, and Other Writings on Media*, ed. by Michael W. Jennings, Brigid Doherty and Thomas Y. Levin, trans. by Edmund Jephcott, Rodney Livingstone, Howard Eiland, 2008 edn (Cambridge, Massachusetts: The Belknap Press of Harvard University Press, 1935).

Bishop, Claire, *Artificial Hells: Participatory Art and the Politics of Spectatorship* (Verso, 2012).

B.O.P. Consulting and Brian Devline Associates, *Arts Council England: Analysis of Theatre in England*, 2016.

Caillois, Roger, *Man, Play and Games*, ed. by Meyer Barash, 2001 edn (Urbana and Chicago: University of Illinois Press, 1961).

Chion, Michel, *Audio-Vision: Sound on Screen*, 2nd edn (Columbia University Press, 1990).

Collins, Karen, *Playing with Sound: A Theory of Interacting with Sound and Music in Video Games* (MIT Press, 2013).

Hill, Tim, Robin Canniford and Giana M. Eckhardt, 'The Roar of the Crowd: How Interaction Ritual Chains Create Social Atmospheres', *Journal of Marketing*, 86. 3 (2022), 121–139, doi:10.1177/00222429211023355.

Huron, D., 'Listening Styles and Listening Strategies', in *The Society for Music Theory*, 2002.

Kaltsa, Vagia, Alexia Briassouli, Ioannis Kompatsiaris, Leontios J. Hadjileontiadis and Michael Gerasimos Strintzis, 'Swarm Intelligence for Detecting Interesting Events in Crowded Environments', *IEEE Transactions on Image Processing*, 24. 7 (2015), 2153–2166, doi:10.1109/TIP.2015.2409559.

Kamp, Michiel, *Four Ways of Hearing Video Game Music* (Cambridge University, 2015).

Kwastek, Katja, *Aesthetics of Interaction in Digital Art* (Cambridge, United States: MIT Press, 2013).

L. Hughes, Roger, 'The Flow of Human Crowds', *Annual Review of Fluid Mechanics*, 35 (2003), 169–182.

The Museum of Modern Art, 'Marina Abramović and Ulay. Imponderabilia. 1977/2010 | MoMA' <https://www.moma.org/audio/playlist/243/3119> [accessed 7 April 2024].

Newzoo, 'How Consumers Engage with Video Games Today: Newzoo's Global Gamer Study 2023', 2023 <https://newzoo.com/resources/trend-reports/global-gamer-study-free-report-2023> [accessed 8 April 2024].

PineII, B. Joseph, and James H. Gilmore, *The Experience Economy* (Brighton, Massachusetts: Harvard Business Review Press, 2019).

Royal Academy of Arts, 'Marina Abramović | Exhibition | Royal Academy of Arts' <https://www.royalacademy.org.uk/exhibition/marina-abramovic> [accessed 7 April 2024].

Secret Cinema, 'Secret Cinema - Terms & Conditions' <https://www.secretcinema.com/terms-conditions> [accessed 29 March 2024].

Small, Christopher, *Musicking: The Meanings of Performance and Listening* (Wesleyan University Press, 1998).

Stanislavski, Konstantin, *An Actor Prepares*, trans. by Elizabeth Reynolds Hapgood, 2008 edn (London: Methuen Drama, 1937).

White, Gareth, 'On Immersive Theatre', *Theatre Research International*, 37. 3 (2012), 221–235.

Wouters, Niels, John Downs, Mitchell Harrop, Travis Cox, Eduardo Oliveira and Sarah Webber, '*Uncovering the Honeypot Effect*', delivered at 'Proceedings of the 2016 ACM Conference on Designing Interactive Systems - DIS '16', 2016, doi:10.1145/2901790.2901796.

Yee, Nick, 'Gaming Motivations Group Into 3 High-Level Clusters', *Quantic Foundry*, 2015 <https://quanticfoundry.com/2015/12/21/map-of-gaming-motivations/> [accessed 12 March 2024].

Zaiontz, Keren, 'Narcissistic Spectatorship in Immersive and One-on-One Performance', *Theatre Journal*, 66. 3 (2014), 405–425.

Media and art examples

Abramovic, Marina and Ulay, Imponderabilia, 1977.

Beyonce *CUFF* IT Dance Challenge #TikTokHypeComps, dir. by TikTok Hype Compilations, 2022 <https://www.youtube.com/watch?v=v-GMPNe-qRE> [accessed 18 March 2024].

BEYONCÉ MUTE CHALLENGE BATTLE - WHO WON? ATLANTA OR NEW ORLEANS, dir. by Foodie in Lagos, 2023 <https://www.youtube.com/watch?v=b6Za yFYSHcs> [accessed 18 March 2024].

Daily Tous Les Jours, *21 Balançoires (21 Swings) | Daily Tous Les Jours* (Montreal, 2011) <https://www.dailytouslesjours.com/en/work/21-swings> [accessed 15 March 2024].

House House, *Untitled Goose Game* (Panic, 2019).

Little Lion Manchester, *The Crystal Maze LIVE Experience in London & Manchester* <https://the-crystal-maze.com/> [accessed 8 April 2024].

Pope, Lucas, *Return of the Obra Dinn* (3909, 2018).

Punchdrunk, *The Burnt City* (Woolwich, 2022).

Umbrellium, *Assemblance* <https://umbrellium.co.uk/projects/assemblance/> [accessed 29 March 2024].

Chapter 3

How do you compose for multiple possibilities?

As previously outlined in chapter 2, audience expectations change when entering an interactive space. As audiences or players are engaging in an exploratory mode they will be expecting elements of interface design through the sound and will read additional layers of meaning into any sound in the space. The successful composer and sound designer must adapt their compositional practices to fit within an expectation that audiences are reading into the sound.

In terms of construction, interactive sound and music provides a unique challenge for composers and the successful composer needs to be able to work in a non-linear format, adapting their knowledge of music composition to provide more flexibility and ensure that all audience members or players have a consistent experience of the music being created.

This chapter will consider the specific challenges of creating interactive sound and music and place them within a contemporary composition framing, building in practical considerations of how a composer can adapt their work to an interactive setting using practical examples and relating these to wider considerations and aesthetic approaches.

The chapter is divided into two sections. The first section addresses sound design and audio as an interface, considering the skills that a sound designer needs to create an immersive world, how the sound can support audience engagement in an interactive work and the implications of audio as an interface to the sound designer. The second section will look at the composition of music for interactive structures considering how this relates to and differs from traditional forms of composition and the different approaches a composer needs to take when developing their work. This provides an outline of both artistic and practical choices that need to be made by composers and sound designers.

Sound design and audio as interface

Immerse and inform

As interactive sound and music relies on audience engagement there are practical considerations to the sound design that relate to the considerations in chapter 2. These include:

DOI: 10.4324/9781003344148-4

- Who are the audience and how much ownership do they have over the space?
- What their expectations are of the work e.g. do they expect free exploration or to be more guided in the space?
- What their understandings are of interactive environments and how will they approach a space?

We can assume that audiences are using different modes of listening in an interactive environment, as previously outlined, the most relevant of these being causal, semantic, reduced and signal listening. Through this, we can also assume that audience members are reading into all aspects of the sound in a place in order to build their sense of narrative and their understanding of how to interact with a space. This means that any sound placed within an interactive space will be approached with the assumption that it is intentional and in some way relates to the immersive environment, narrative or guidance for interaction.

With these assumptions in mind, sound designers in an interactive space need to be extremely intentional with their work.

As a structure for understanding, I would divide all sound in an immersive and interactive setting into the following two categories:

- **Immerse** – Sound that embeds you in the world and the narrative
- **Inform** – Sound that tells you how to interact with the world

This provides a useful framework for sound designers developing their work, and a framework for analysis of sound design in an interactive space considering both the creative and practice elements of the work.

Sound to immerse is anything that is diegetic to the space or sound and music that supports the emotional narrative in a scene. This can be:

- Atmospheric sound that gives a sense of the place, e.g. world building sounds that define the environment such as creature sound, weather or atmospheric sounds such as vehicles
- Sound that comes from objects in a scene such as radios or technology
- Non-diegetic musical scores that build the emotion in a scene
- Atmospheric sounds can also support the emotion in the scene, such as bass heavy sounds that build a sense of dread[1]

Where the term immerse has been used this is in a narrative sense related to world-building in an immersive setting and is not related to audio protocols or binaural sound.

Sound to inform is any sound that can be considered part of the interface or to build the player or audience's understanding of the space they are in. This can be:

- Sounds that guide players through the space, these sounds tend to be higher frequency locational sounds with little or no bass

- Audio used as an interface that gives players or audiences an understanding of how to interact with the space
- Audio changes that give an indication of player performance such as live filtering used as an indication of player health
- Elements of the narrative including fragments of speech intended to guide a person through a space, or radio segments that give an indication of what is expected of an audience member or player

There can be overlaps between immerse and inform as structures. For example, in *FIFA 1998*[2] the commentary from real life pundits keeps you immersed in the game but also gives an indication of what is required from the player who can use signal listening to time their strike with the clues from the commentary. Equally, a non-diegetic score may be used to inform a player of a change in gameplay or an impending fight.

Audio as an interface

A key function within sound to inform is the role of audio as an interface. This is audio that supports an audience member to understand and interact with an interactive piece of work.

Jakob Nielsen has developed heuristics for User Interface Design that usually apply to visual interfaces within a computer design framing.[3] I would suggest that these can be further expanded to apply to immersive theatre and other interactive environments where the audio is used to support an audience's understanding of an environment or expected interaction.[4]

Within Nielsen's ten heuristics for User Interface Design, I would consider six applicable for an interactive audio environment.

Visibility of system status: This particularly applies to a game environment where audio can inform players of how well they are doing within the game without them having to look at a performance bar. For example, in a game such as *Call of Duty* when a player is close to dying, dynamic high cut filtering is used to remove detail from the sound in the scene indicating the system status.

Match between system and the real world: In an immersive environment, when sound to inform is used it needs to either follow real world examples and standards or have its own internal logic of how the sound works For example, a naturalistic immersive theatre production looking to represent an air-raid siren to evacuate audience members to a different space should use a historically accurate air-raid siren sound. Similarly, a children's production based on a television show would want to use the sound effects that the children regularly hear when watching the show.

Consistency and standards: Sound within an interactive environment must be consistent in order to support interaction. The extent to which consistency is needed is dependent on the genre, setting and the match between the system and the real world as outlined above. For example, in a platformer game, such as the *Super Mario* series, consistency is vital to reflecting the genre and support-ing the learning of the game. When Mario jumps, you hear a jump sound, and this is repeated however often you jump in the game. This supports you to connect the button that is being pressed with the action of jumping.[5] In a more realistic or immersive space the consistency may be expected but with a number of variations that reflect the real-world experience in a space. For example, the sound of an engine starting in a car is consistent but is not identical every time.

Error prevention: Audio should support the audience to understand how to interact with the world. For example, it should support movement around the space or indicate how to interact with an object without the need for further signage or instrument.

Aesthetic and minimalist design: All sound within the work must serve the immersion within the narrative or installation or provide a practical function to support the audience's understanding of and interaction with the piece. As we are working from an assumption from chapter 2 that audience members will read into all elements of the sound as being intentional, no sound should be included that does not serve the function of the work.

Help users recognize, diagnose and recover from errors: As with error pre-vention, sound has a practical role to support the audience's interaction with the work. Error sounds must be consistent, as outlined previously, and fit within the inner world of the game.

These heuristics of interface design are used to underpin the principles below when considering sound to inform.

Unpredictable audiences and audience behaviour

The biggest challenge when writing interactive sound and music is the amount of control that the composer or sound designer has over the finished piece of work contrasted with the level of interaction and perceived interaction.[6] Once your work is released into the world you will not necessarily have full control over how it is interacted with or the final outcomes.

If you're considering an exhibition or installation space, audiences will:

- Move through quickly as they're looking to get to another gallery/experience
- Try to explore the entire space collecting all the experiences available within the installations
- Try to see the seams of the installation by testing it to its limit – this could be trying to wait for the ends of loops and repeatedly testing functions

Understanding audience behaviours and expectations, as outlined in chapter 2, becomes a great skill when working in an interactive environment. Tools such as player testing or working with a test audience can help understand a sound designer's understanding of how people will act within their work, but the work will be at a late stage of development by this point and any changes will be edits/updates on the original design. Further to this, test audiences can often be 'friendly fire' made up of friends and family who want to support a project and have heard you discuss the work in development or play testers who are experienced gamers and know how to approach the interactive space. Many installations fall foul of this mistake, and you will see this demonstrated through detailed instructions of 'how to interact' or gallery staff instructing you on how to get the most out of the piece.

As outlined in chapter 2, audiences arrive in an interactive space with a large amount of expectations but their behaviours are also impacted by the use of sound and music within a space.

Practically speaking, as a composer or sound designer working in an interactive environment you may need to work with directors on elements that support narrative or audience flow.

A unique challenge to immersive theatre is the shift in audience expectations. Within typical theatre audiences there are rituals of theatre and audience expectations. Traditional theatre attendance is a highly ritualised experience which relies on the audience member taking an observational role – 'bearing witness' to the action. Within a theatre production there are three distinct environments: the foyer where you meet prior to the performance where programmes and drinks can be bought, the time spent in-seat pre- and post-performance and in-seat once the curtain rises. There are preparatory signals to the audience members to begin to adjust their behaviours – the bell ringing, and lights dimming signalling five minutes prior to the show starting and (in more recent performances) the announcement to switch off mobile phones.

In contrast, a gamer who regularly plays open-world games will have a set number of actions that they have developed to support their investigation of the gaming world. For example, you look behind every waterfall, you test every door and you try to talk to everyone who might be able to offer you help with completing your quests. This means that if we are building interactive sound for a non-gamer audience the impetus to interact with the space will be different.

A further challenge with immersive theatre that will impact both traditional theatre audiences and game literate audiences is the implied theatrical construct of 'sharing with strangers' – Christopher Small discusses this in *Musicking* by saying:

> …most, if not all, of the audience will be strangers to us. We are prepared to laugh, to weep, to shudder, to be excited, or to be moved to the depth of our being, all in the company of people the majority of whom we have never seen before, to whom we shall probably address not a word or a gesture, and whom we shall in all probability never see again.[7]

In a traditional theatre or game environment, we have an additional artifice available to shield us from interaction with strangers, for example we're sat in the dark facing the stage or we're talking over an online communication system which allows us to detach slightly and skip any small talk. This extra protection is gone in immersive theatre where you are confronted with the stranger and asked to interact directly with them. In essence we are moving what has become an individual experience, shared within the same place or remotely online, to something more collaborative where communication with strangers is a necessity.

All of these production types within immersive theatre rely on the audience being able to have individualised experiences, for example a one-to-one inter-action with a character in the play or finding a secret room within the set. These can be used as a promotional tool for the show, with audience members sharing their experiences online via social media. However, this can only be an effective tool if the audience are able to feel like they have found the experience through their own choices.

Sometimes though, audiences and gamers can just be unpredictable. Within installations and shows I've had players:

- Lie down in the middle of the gallery so that they can 'spot the end of the loop'
- Deliberately do the route around the gallery back to front to see if that changes the experience
- Try to get behind exhibits to see how they're wired
- Try to trigger sounds multiple times in quick succession to see how much sound is in the installation and if they can break it

While not the expected approach to the work, this doesn't mean that the installations were failures but that people were experiencing the work in the way that suited their needs.

An aim of the audio should be to encourage intuitive interaction in a space, guiding the audience as needed while building on the role of audio as an inter-face as outlined in chapter 2. This is not at odds with such unpredictable audience behaviour.

Quality assurance and playtesting approaches

Sound designers and composers working in an interactive space need to be aware of the unpredictable approaches different audience members will use and be able to effectively use tools such as Quantic Foundry's gamer motivations,[8] which highlight the different ways that a player may approach a space, to test their work almost to a breaking point; this builds on quality assurance roles in games and requires the composer or sound designer to interact with their work in multiple ways.

First is the practical function of whether the game or interactive installation is functioning as expected. This includes:

- Listening to any looping ambient sound and music, is the end of the loop audible, are there any audible clicks or pops? When transferring a looping sound into a game engine or a larger ambient space such as gallery these loop ends can become particularly noticeable and may impact the audience's experience of the work

- If there are boundary triggers to bring in the audio (e.g. trigger boxes in games), walk across them multiple times. Does the sound trigger as expected and is this consistent? If the sound is meant to only trigger once does this occur?
- Checking if triggers can accidentally be avoided by the player e.g. can they walk around a trigger or jump over it?
- Is the balance of the sound in the scene as expected and can audio interface objects be heard above the general soundscape?
- Consider what your expected audience numbers are and how this will impact the interaction and the effectiveness of the technology involved. Is there a saturation point where the technology will stop working?

Then the sound designer should start testing for different player behaviours based on player motivations.

Speed running players/audience members:

- Does the sound have as much impact when heard in a short burst or is the larger structure of the work lost in a speed run?
- Is interaction instant or does it take time to start a process? In this case could an audience member misinterpret the lag as the interaction being broken and try to start the process again?
- How does the system cope with repeated starts to the audio, does this cause audio glitches?

Exploratory players/audience members:

- Are there any distinctive sounds in the atmospheric soundscape that would draw a listener's attention and show that the sound had looped?
- How long is the audio loop, is it possible for an audience member to stay for multiple loops of the same audio?
- Are there any repeating sounds or music that would highlight to a player or audience member that they have spent longer than expected in a section? Is this intentional or could this impact audience experience?
- Is the sound consistent across all elements of the space? For example, is it clear to the audience which objects belong within the narrative and do these objects all have sound consistently applied if an audience member chose to explore additional objects and sound sources outside of the main narrative or focus of the installation?

This playtesting approach requires the sound designer to have an eye on both the practical and aesthetic functions of the work and requires them to use an imaginative audience-centred focus to consider how their work will be experienced. Through taking this time in testing phases they are able to predict audience behaviours and build a stronger, more considered audience experience.

Narrative fragmentation and guiding audiences

The risk of interactive experiences, particularly within a narrative framing such as narrative games or immersive theatre, is that audience members can lose out on key pieces of information that can help them to understand the work as a whole. When audiences have explorative control where they control their pace through the set, for example in an open world game, a linear narrative can become fragmented leading to audiences feeling like they have missed out on key experiences and make them feel disorientated. For example, in Punchdrunk's production of *The Duchess of Malfi*, an opera staged where the audience have exploratory control, *The Guardian* noted:

> The essence of the experience is randomness: it's rather like the William Burroughs technique of arbitrarily re-arranging sentences to defamiliarise them. But the problem with applying this to Webster's play is that it assumes you know the story in the first place: if you don't, you'll be hopelessly lost. The approach also works against the cumulative impact of any piece of music-theatre: it becomes difficult to judge Rasch's score since you only hear it in disjointed fragments, from small woodwind or string-and-brass ensembles, until the final scene. It essentially becomes background music to a series of Websterian verbal highlights.[9]

This can lead to feelings of disappointment within audiences as highlighted by Atkinson and Kennedy's research into Secret Cinema events which draws attention to the gap between audience expectations of the event and the fragmented experience that can be presented during busier performances.[10]

This fragmentation can also happen in games where unstructured exploration can impact a player's understanding of the narrative, or unlocking of elements of gameplay for example, on a play of *Batman: Arkham Asylum*[11] by following arrows I turned right instead of left, meaning that I didn't trigger the main mission and spent time exploring empty rooms.

There are some very simple solutions to narrative fragmentation that can lie outside the use of audio. The first of these is signposting, having arrows or similar that physically point the audience to where the next scene or information is being held. Secondly, actors can be used to usher the audience to the locations by a physical herding effect or a more subtle swarming, where if enough people from the cast move then the audience will join. Finally, certain routes for the audience can be closed off to limit the amount of exploration that they have access to. Doors can be locked and paths can be blocked to limit them to one correct pathway. All of these solutions are less than ideal, as they remove the element of choice and freedom that makes immersive theatre so appealing to audiences. Similar to when a game limits your interactions with an NPC to only include a few, frustrating, pre-set options, for all the solutions above audiences have been given the illusion of control when this still lies entirely with the director. More elegant solutions can lie in the strategic use of audio and modes of listening within an interactive space, within the functions of sound to inform. As Collins states:

> We also can extend signal listening beyond music by listening to sound effects for navigational information about direction, proximity, and spatial cues; status information about a process or event; and semiotic information on the nature of virtual characters that we meet or places that we encounter.[12]

This can be done through locational audio within sound to inform; building on understanding of how people engage with sound on a day-to-day basis.

Higher frequency sounds are highly directional and people engaging in causal listening are drawn to them, e.g. consider the frequency of a siren or a car alarm that allows you to place the sound within a location. A high frequency sound can be placed behind a door or in the location where you want to move your audience and the people will move in the direction of the sound.

It should be noted that high frequency sounds lose their effectiveness over time, especially if one static note or chord is used to guide people. Humans are constantly filtering out unwanted information and static sounds become almost a wallpaper after a certain point, the audience is aware of them but has stopped paying attention (e.g. the smoke alarm requiring a new battery that beeps regularly that some people, not all, are able to ignore after a number of days).

The trick with ensuring that a locational higher pitched sound keeps its guiding function is to keep it moving. This can be done gently be creating what I call a 'sparkly sound'.

A 'sparkly sound' is a higher frequency sound with an automated filter applied that keeps that sound shimmering, allowing you to keep the sound moving in an area, gently guiding the audience to a space. It continues to catch the attention and more detail in the sound is revealed as you move close to the source, rather than becoming part of the general atmosphere.

Another technique is to build on the use of the narrative to bring people into a space through playing snippets of speech, e.g. in a video game where you can hear a character calling for help from off screen. People are drawn to speech due to semantic listening. When we hear speech in an otherwise quiet area we try to find the speaker and determine the meaning of what we are hearing. Hiding speech within a space will naturally guide an audience to try to build the narrative moving them through the space, this is a common technique in narrative video games such as *Batman: Arkham Asylum* where the player needs to follow cues such as the Joker's taunting to find his location and advance the narrative.[13]

Giving user feedback and semiotic associations

As audio acts as an interface for audiences in an interactive space, one of the functions of sound to inform is to give user feedback. This is an expectation for game literate audiences but is very helpful when encouraging interaction from a non-gamer audience. Audio becomes a key tool in encouraging interaction and in error prevention.

Audio as a feedback tool can be built into the interface, for example if there is a puzzle in an interactive space the sound should give an indication of whether the puzzle has been correctly completed or not. Sound can also be used to give instant feedback of whether an action has been selected or a button has been fully pressed. This is important when working with audiences as instant feedback ensures that audience members do not continuously restart an interactive process as a result of not knowing if their action has registered.[14]

If not using a clear interface, for example when working in an immersive theatre setting, feedback sounds linked to the narrative can also be embedded in a 'talisman object', which is a prop or environmental object of importance to the overall game narrative. Here a sound can indicate the importance of the object to the narrative or a character, or indicate to an audience member that they are following the correct path.

As audio for feedback is a tool to inform audiences of whether they have interacted 'correctly' with the work to be effective as a form of interface design, as previously outlined, it needs to be consistent and, where appropriate, match real world standards. This interface design requires a sound designer to have a strong knowledge of semiotic associations as this will impact the reaction that an audience member or a player has to a sound. For some designers this may be unconscious or embodied knowledge built from their own consumption of media.

Semiotics, the study of signs and symbols, builds on learned associations that may have developed over hundreds of years using examples from the church, opera or other musical forms. For example, a rising perfect fifth played using brass is often used to represent a heroic character, this builds on operatic associations and can be heard across the work of film composer John Williams.

A bell-like 'ding' often has positive associations with gaining points in a game, linking back to more physical game sounds such as pinball machines. This bell is also a common text tone alert on mobile phones, which may then change the cultural associations.

Symbolic associations are region specific and do change based on country and other cultural considerations and may not translate across territories. The examples above have been written from a Western classical music perspective.

Very practically within the use of audio as an interface, a sound designer will also want to build a suite of sounds to connect different interactions and their outputs, e.g. an on and off sound will be the same instrument and timbre placed an octave apart (higher sound for on), so that the audience can see directly how those two actions are related. This, again, builds on semiotic associations and also allows for consistency of use within the interface.

Talisman objects will need to use more complex sounds to indicate how they relate to a narrative. This can include use of leitmotif, building on opera traditions, to associate an object with a character in the narrative, or sound and music related to a place or feeling (e.g. a chord built around dissonant frequencies to link an object to a sense of dread).

Having an understanding of symbolic associations with sound provides a practical way for a sound designer to implement audio as an interface, based on Nielsen's Heuristics. Audience members can use their existing associations and don't need to build a new series of associations in order to interact with the work. It also ensures that the sound designer does not fall into traps where they have added a sound that they think is interesting but its inclusion leads audience members to infer additional detail unintentionally impacting their engagement and interaction with the work.

Narrative immersion and environment

Returning to sound to immerse the audience or player, there are specific concerns for the sound designer when creating a physical environment for an interactive work.

In an immersive interactive environment, such as an immersive theatre production or an open world game, a full world needs to be built to support the player or audience's understanding of and immersion in the space. In an immersive theatre production the sound designer can either work with or against the sounds from the outside world, choosing to incorporate the outside world into the soundscape, highlighting key elements, or choosing to completely block out the sound by building a new world soundscape from scratch for the ultimate escapism. In a game context, the sound designer is responsible for all elements of the sound world and how this relates to the wider sense of the environment. Both of these environments require an understanding of 3D sound in a way that is not necessarily required in other media.

For both soundscapes we must consider the audio from the perspective of the audience member or player in the space. In games the sound is heard from the perspective of an audio listener object, this is usually placed on the camera so all sound is heard from the perspective of the player and reacts according to the player's location and viewpoint (in a third person game this may be different relative to the perspective of the characters).

In an immersive theatre production this is similar, as audience members walk around the space they will experience the sound relative to their position in the room.

Both environments require the sound designer to have a working understanding of how sound reacts and environmental sound including physical properties of sound.

Sound designers need to have a physical real world understanding of sound behaviours including how different sound waves react, and sound roll off in different environments. They need to understand elements such as occlusion and how speaker placement and sound location can affect how a sound is heard in a space.

First the sound designer needs to build the inner logic of the environment, which includes working collaboratively with other departments on a project and includes answering the following questions:

- What is the aim of the work? Are the team aiming for complete escapism/immersion or is it acceptable for the real world to be heard within the space? This can impact the types of sounds chosen and speaker placement within the work
- How should people feel when they walk into the room? For example, should there be a feeling of dread in the room or is it a welcoming environment?
- What is the setting for the environment? A built-up environment has very different needs to a more rural environment. Similarly, a historic environment may dictate that the modern world shouldn't be audible in the work
- Are there any creatures in the space? Do they relate to real world creatures?
- How long are people expected to stay in the space?

Any decisions need to be consistent for the entire scene so that the audience or player can be fully immersed in the environment. While not linked to the interface design, the principle of consistency and real-world match still stand to ensure the audience is able to immerse themselves in the environment that has been created.

To build in realism within an audio environment, the sound designer needs to have an awareness of roll off and 3D sound. In a real environment sound happens at different heights and there are sounds in a room that we may not always be aware of, due to the ability that many humans have to filter out unwanted information. These hidden atmospheric sounds can be found in everyday settings by leaving a mic recording in a seemingly silent room, e.g. in the kitchen you'll find that your fridge is much noisier than first anticipated and other electronic devices have audible frequencies. This is the detail that is required when building an entirely artificial audio environment, e.g. a game environment, to give the soundscape a believable depth. In a theatre or gallery environment this is less of a requirement as the room the work is being staged in will have its own characteristics that the sound designer can work with and augment.

Further to this, a sound designer must understand the acoustics of the space that they are working in (virtual or real world) as adding sounds with applied reverb, or reverb from the space in which they were recorded, may make the environmental audio sound inconsistent with the space that they have been placed in. This can create a conflict with the audience or player that will disrupt any sense of immersion. In game settings it is preferable to add the reverb to dry sounds through middleware processes, using reverb zones, so that the sound will be realistic to the space.

For a real world setting the designer must first build an understanding of the space and may wish to use convolution reverb for the space that they are working in. They will also need to consider the relative loudness of sounds that they are adding to the soundscape. For example, if it is an environmental sound being played in through a speaker, is it localised to an area? Does the sound level sound realistic to space? To emulate this, I often forgo expensive speaker systems in favour of smaller, less powerful 'drop' speakers that I can hide in and behind objects so that the sound is more natural to a space. Here realism is preferable to audio fidelity.

A very practical consideration for audio world-building in an interactive and immersive setting, is the length and features of an audio loop. If a loop is too short the same sound will be heard multiple times by the audience, this will reduce the effectiveness of the atmospheric sound as a tool to support the narrative and sense of place. This is particularly important if very distinctive sounds have been used in a loop (e.g. certain types of bird song, alarm sounds, speech etc.). Humans are built to recognise patterns and after hearing a sound twice audiences will begin to associate it with the repetition in the loop. This can affect enjoyment of a scene as they may feel that they have spent longer than expected in a scene or that they are being rushed on to the next section.[15] To address this the sound designer may need to build a soundscape with fewer distinctive elements so that there is less that an audience can recognise in the loop. This does take away from the specificity of a soundscape and can make the sound world a little generic.

Another approach for the sound designer is to build a longer audio loop. This requires the designer to consider how long they want the audience to stay in a scene and how long members with different player motivations may stay in a work if they are exploring the space. Practically speaking, when building a soundscape, I make my audio three times the length that I think people will be staying in a scene, which does require a lot of audio and large file sizes but provides additional space for people wanting to explore a scene at length.

Finally, a practical solution that also saves on file size and storage is to use a randomised approach to building soundscapes. This can be easily accommodated using game engines or software such as Max. This requires the sound designer to consider their work in terms of the prominence in the soundscape, where some sounds will be continuous and others will be triggered at random to bring interest to the soundscape.

Table 3.1 Structure for atmospheric loop design

Atmospheric loop design
Background sounds e.g. sound of weather, background hums, atmospheric rumbles Triggers once and loops continuously
Mid-level sounds e.g. traffic Runs continuously with a different variation triggered on each loop
Top-level sounds e.g. bird call, tech alerts Triggers at random with multiple variations included

Sound designers would need to be aware of how they programme randomly triggered audio as a computer will consider this random and may trigger the same sample multiple times in a row, which conversely will be perceived by audience members as not actually being random (even though it is) due to the perceived pattern in the sound. To address this the sound designer can put conditions on the randomness about how soon a sample can be used after the original trigger or to ensure each sample is used once before they can be repeated.

Through this understanding of the physical features of real world audio and human tendency for pattern recognition, the successful sound designer is able to build a realistic and consistent sound world that will support the role of sound to immerse.

Composing for interactive environments

Composing for an interactive environment requires a combination of skills, building on traditional composition knowledge from harmony and counterpoint practices through techniques developed in 20th-century experimental sound and music. A successful composer in this field will have a range of techniques that they can adapt based on the needs of the medium that they are working within.

Within music writing, structures have been developed whereby the composer's intention for a work is the most important facet, this builds on understandings from more classical context built on the concept of the musical work.[16]

However, when the composer's intention is to create something interactive, that changes the compositional process as the composer's control comes up against the unpredictability of humans.

With a truly interactive space, from a composer's perspective you have to be prepared for your audience to enter the space or leave it at any given point in the work. It has to exist within a structure that is not reliant on linearity.[17]

Within this section I have made a distinction between interactive sound and music that adheres to linear forms and those with slightly more freedom. The structures provided for composition techniques will move from a linear approach with some flexible structures through to a completely interactive form.

The phase space in interactive music

Composing in an interactive environment requires a different understanding of the compositional process.

When Philip Pullman discusses writing fiction he describes a phase space, an idea repurposed from physics.[18] This is a space where all the possibilities exist for the narrative at the same time and they all exist with equal weighting or likelihood of happening. As the author makes their decisions, some routes through the space become more prominent or likely as they build from previous decisions. Within Pullman's discussion of this phase space the other options for the narrative remain like shadows of what could have been in the narrative, like a parallel universe of possibilities.

> Every sentence I write is surrounded by the ghosts of the sentences I could have written at that point, but chose not to.[19]

This is similar to writing music. When you start writing a piece you have a range of choices or limitations that you make. You choose your instruments, key (or lack of), time signature (or lack of) and an outline of structure for the work. As you progress through the composition process you make decisions to work with or against the expectations. Should you choose the expected chord progression or work with something different? As you move through the composition process the choices become more expected and more refined, solidifying the work in a form.

Through this, you construct the piece, but the other options remain as choices that were not taken.

Interactive music requires a less solidified approach, leaving the options open. As a creator of interactive audio, you are required to see the phase space. Instead of designing one clear path through the options you have to think of all the possibilities based on an unpredictable unknown audience. Rather than looking at one linear piece of work, you have to look at possibilities that branch out and the consequences of not just your decisions but the audience's. Essentially, we are keeping part of the phase space alive. These are not the ghosts of sentences that haunt the work with prior possibilities. These living possibilities in the music are waiting to be chosen on any given day. They're like a substitutes bench of music, a squad of players that will suit different needs and will wait for the right moment to be deployed into the game.

This requires different skills from a composer who is required to not just follow through one musical path but to musically imagine multiple possibilities, sometimes of so large a combination that they won't hear all the finished results.

Structuring linear composition

Structures of composition are often based on linear concepts related to time. From harmony and counterpoint to larger sonatas across to pop song structures, we move through the work along a timeline. Similar to narrative design in media, the most common exception to this structuring is media such as video games.

Audio in narrative-driven video games or immersive theatre contexts can still be considered linear in form with some inbuilt indeterminacy. Players are working through levels, stages or open worlds but they still often have a narrative arc that they are expecting to experience with a clear start and end.[20]

For example, in an open world game such as *Legend of Zelda: Breath of the Wild*,[21] which is notable for having a lot of open world exploration, you still have the following linear functions:

- Transitions from day to night
- NPCs (non-playable characters) interacting with Link (the main playable character) and providing side quests

- Fights
- Temples with puzzles
- Triggered cut scenes
- The overall game narrative

So while the world is open, it is still very much time bound. Crucially, in addition, the audience member or player is expected to be there for the entire duration of the game. The player is the driving force for the narrative and the narrative cannot exist without the player. Furthermore, it is expected that you will have continuity in the player experiencing the narrative; it is not expected that you would start a game with one player and then transition to other players to complete the game with the original player never returning.[22] This is due to the nature of control in video games and how they are built. The player is, generally, the focus and the world is built for them to be experienced from their perspective (even if playing a third person game).

Due to this narrative or linear function it's important that the sound matches the action on screen and the narrative needs of the scene. In the same way that audio for a film would be developed, you are supporting the narrative or linear arc of the scene. The indeterminacy in linear forms of interactive sound and music is to do with how a player approaches a game, how long they spend exploring and how much action they bring to their approach.[23]

The challenge of a fully linear structure is that, traditionally, it pushes elements such as build and release, time signatures, key signatures etc. These tools are in place to support us to structure composition and to choose a route through the phase space. By choosing an overall structure and shape to the work, you reduce the potential paths through. Each choice to limit makes the composition itself easier to complete. In an interactive space, these spaces do not just limit us, they also limit our audience. For example, the choice to build the tempo pushes our audience member towards decisions and influences their gameplay. Similarly, traditional build and release, where the music builds to a peak of energy, pushes our player or audience to reach conclusions about the narrative.

However, there are lessons that can be taken from more experimental 20th century forms, such as minimalism, that support the needs of interactive music to provide flexibility while supporting the player or audience to retain control over their experience and pacing.

Loops and borrowing from minimalism

The most basic way to build interaction into your work is through use of loops and layers.

Loops and layers use an indeterminacy approach to constructing music while still retaining a linear structure to the music. Here, your player or participant has some control over their pacing but will eventually engage with the narrative in the order that you are intending.

This is a block-based approach to composition where the composer builds the music through repeating sections that can be combined through a theme. For example, the music in a game such as *Hades*[24] is built using a number of different themes. The music starts as you enter a chamber and then the central themes loop for as long as you are playing the level with some additional licks or motifs added during game play. This means that the music can continue throughout the rooms, adjusting in length to how quickly you clear the number of enemies without creating a sense that you have been in a room or level for too long.

A loop-based approach has an impact on the harmonic rhythm of a work as in order to build flexibility with the repeating blocks and ensure that the music is able to loop seamlessly, the blocks need to start and end on in the same key, leaving little space for modulation.

Based on this the harmonies in the level stay quite static over a fast-moving beat. This means that you will have a slow harmonic rhythm over a (potentially) faster rhythmic tempo which can create a sense of stability among the movement.

The start and end of repeating blocks also become important in looping structures. In order to signal the end of a fight (all enemies cleared), the music has to have a clear outro.[25]

This means that the structure of a looping track is as follows.

Table 3.2 Basic looping structure

A	B (Loop)	C
Intro	Looping main section (transition based on bar ends)	Outro

There is a risk when repeating sections that the music can become overly repetitive and this can impact the player or audience's experiences of the game or installation. For example, if you are stuck on a level or within a puzzle and the same musical motif keeps repeating this can compound the feeling of being stuck in a level. A composer can address this by adding more variation through having multiple melodic blocks that can trigger at random[26] giving this structure:

Table 3.3 Outline of looping structure

A	B (Loop)	C
Intro	Loop 1	Outro
	Loop 2	
	Loop 3	
	Loop 4	

These loops can be of an irregular length but will generally need to be of a consistent tempo and key.

From a composer's perspective this has implications for the structure of a melody. Generally, we consider the rise and fall of a melody similar to that of sentence construction, you consider where the melody is headed in an overall structure and the musical sense of that construction.

Similarly, the music has to remain in a consistent key as the repeated blocks of music are being used. This, combined with the slow changing harmonic rhythm, changes how a composer constructs a melody and can seem restrictive to a composer. However, it is still possible to include a wide range of variation in a looping block interactive composition.

Melodically, working with interchangeable blocks borrows compositionally from minimalism in an indeterminate form. A good comparison with this musical structuring is Terry Riley's *In C* (1966);[27] this is a large-scale work of indeterminate length built out of melodic blocks over a repeating metronomic C. The performers repeat each block as often as they would like before moving onto another section.

There are key changes within *In C* but only within related keys, e.g. block 14 introduces an F# signalling a move to the dominant. There is rhythmic and harmonic complexity and the melodic material is built with different bar lengths (section 1 is 3/4 while section 35 is 30/4), although due to the nature of the work these sections can be interpreted differently through listening. With these differences, the melodic material is built in a way that every adjacent block works as a melody. However, one could argue that while the music is indeterminate the piece is still constructed linearly.

Vertical re-orchestration and borrowing from harmony and counterpoint

While looping is an effective tool for dealing with indeterminacy in interactive media, it does not fully consider the musical needs of a larger or more complex narrative.

Consider this in a similar framing to large AAA games such as *Star Wars Jedi: Fallen Order*.[28] This is a narrative game with a clear plot. The player has the ability to explore the planets but, essentially, you are working towards a series of set pieces (battles, puzzles etc.) that all build to the overall plot. The way that the player approaches the game will ultimately decide the pace at which they meet each of these set pieces but not the overall narrative structure. In a game this large with so many spaces for exploration, the music needs to respond to the emotional needs of the game but also to how actively and aggressively the player approaches the game, e.g. the player can choose to engage in all battles that they reach or can take a more stealth approach avoiding fighting. This requires a more adaptive mode of writing that allows different levels of musical intensity based on narrative need and player actions.

As a composer, to develop the block approach you need to build some variation into the loops. This can be done through composing interchangeable melodic and rhythmic blocks built up into layers.

To successfully create layers in composition, allowing for flexibility in orchestration, a technique called vertical re-orchestration is used.

In vertical re-orchestration a fully orchestrated version of multi-instrumental music is written, including looping blocks. Each instrument is assigned to a fader on a mixer and will be brought in and out within the mix to increase or lessen the intensity of the music as needed. These blocks can be combined in any order based on the needs of the scene and the player's actions. For example, you could have blocks of chords playing on a repeating loop that provide a passive version of the music which then builds in intensity by the addition of a bass line, followed by layering a beat.

The benefit of this technique is that it allows for indeterminate structures while supporting the narrative needs of the scene and player decisions. Rather than providing multiple musical cues for a scene, the music can transition seamlessly through the live mixing process.

When exporting stems to assign to faders, vertical re-orchestration requires the composer to consider the groups based on intensity level rather than based on usual instrumental pairings.

For example, take this repeating ten bar piece:

Figure 3.1 Looping ten bar orchestral music, L.A. Harrison

Rather than exporting by instrument family, we consider the lines by intensity and musical function, ensuring that each section also makes musical sense without the other lines being included.

Violin I, II and viola comprise the basic calm state for the work, noting that they still retain some melodic function. The oboe provides the main melody with flute added for countermelody where interest is required, the percussion and timpani add intensity and the highest intensity is achieved with the French horn, cello and contrabass.

All these layers can exist independently, e.g. the oboe can play solo without the underlying strings.

When combined into a looping structure this looks like the following:

Table 3.4 Outline of looping structure for vertical re-orchestration

A Intro	B (Loop)	C Outro
Intensity 1: Violin I, II and viola		
Intensity 2: Add oboe		
Intensity 2: Add flute counter melody		
Intensity 4: Add percussion and timpani		
Intensity 5: Add French horn, cello and contrabass		

This can be further augmented by adding variation to melodic material in each block so each line has melodic changes.

In the construction of the melodic material, vertical re-orchestration borrows from the technique of voice leading in harmony and counterpoint where the composer is required to think both vertically and horizontally. All vertical lines of music must make harmonic sense while all horizontal instrumental lines must made melodic sense (or, in the case of percussion, sense as an individual line). This provides the composer with flexibility as each instrument (or group of instruments) can be removed or added in different combinations to provide more or less intense versions as needed.

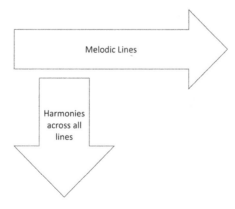

Figure 3.2 Requirements for voice leading

For the maximum number of combinations of intensities, the most practical way to develop this as a composer is to think both about the desired intensity of the music with different forms of gameplay and the sing-ability and interest level of each of the instrumental lines horizontally. For example, in Figure 3.1 the violins and violas all contain melodic material, while the rhythmic structure for the percussion, French horns and lower strings is strong enough to retain interest in a loop before other layers are added.

Non-linear structures

Loops and layers work well as compositional technique when there is a narrative structure supporting a linear format to the work. However, for a sound installation the requirements of the work are different. In an installation that runs continuously in a gallery throughout a day, there is no clear start and end to the work. Audience members will enter and leave throughout the piece and will stay for different durations based on their engagement with the piece.

While loops and layers can work for this format, the linear structuring of music can lead to the perception of a start and end of a work and audience members feeling that they have missed out on part of the experience, leading to dissatisfaction. The ideal for installations is that audience members have the same quality of experience whether they stay for 30 seconds or for 20 minutes.

Moment form

As most composition is based on linear construction, form provides a challenge for interactive installations as an audience member can enter and exit at any point in the work.

While in composition we focus on linearity in our structure (A leads to B leads to C etc.) which can include the build and release of the work, compositionally interactive works and compositions can become fragmented and lose their impact if not heard in their totality.

One solution to this is the use of moment form, as developed by Stockhausen in his work *Momente*,[29] which provides the ideal solution to some of the challenges presented in structuring interactive music.

Moment form is built on the principle that every moment in the work is of equal importance to the one that came before and the one that will come after. A moment can be a small slice of the work or can be the piece in its entirety. Each section, however it is sliced, is of equal importance and contains its own inner sense of structure.

This is ideal when building a sound installation for a gallery or a public art context. If the composer is aiming for each audience member to have an individualised experience and an experience of equal worth, working with moment form removes the need for a traditional start and end to the piece and can be used effectively with more acousmatic soundscapes and looping structures.

Moment form requires the composer to work in both the micro and the macro form. When composing the piece they need to consider the overall macro structure and the structure of multiple slices of music, slicing the work into almost infinite pieces.

As can be seen from the figures 3.3 and 3.4, zooming out shows an overall structure of the work, while focusing on different sections of the piece demonstrates smaller microstructures in the work.

In a way, the composer is aiming to construct a fractal-like structure.

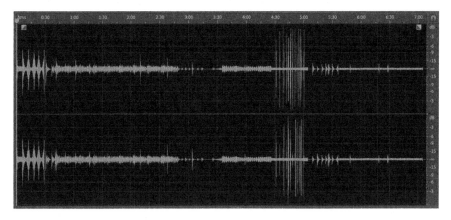

Figure 3.3 Full structure of piece written in moment form: KA, L.A. Harrison, 2012

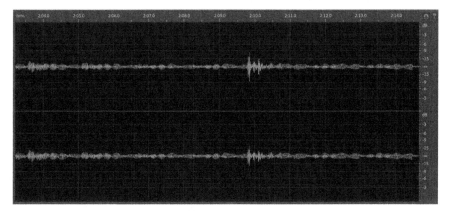

Figure 3.4 KA, L.A. Harrison, 2012, zoomed to show structure at 2 mins 3 secs – 2 mins 14 secs

This requires the composer to always have an ear to the audience experience of the work, which can be achieved through the editing process rather than through the overall construction of the work, especially due to the lengths of pieces being created.[30] When constructing a composer may be working moment to moment, but this can be refined through the editing process; I think of this as a sculpting process, finding the piece in the constructed material. Using moment form, the composer has to put themselves in the position of the audience, which means listening in different ways, similar to the quality assurance-based play testing approach outlined earlier. First they listen to the work as whole to ensure that the overall structure is satisfying and meets the compositional needs, then to moments of different sizes of random to check the structure of each moment and ensure that each provides an equal experience to the audience member.

Moving away from pre-determined structures

All solutions given so far for composition have supposed a structure, linking to traditional or experimental composition processes. An alternative for truly interactive sound and music that is satisfying in a gallery or public art context is to move away from predetermined structures completely, providing the audience with the most control over the outcomes of the work. This involves having no pre-written material within the work and can be generative in form or completely improvisatory. This links to more approaches from indeterminacy similar to those developed by John Cage and Morton Feldman where entire structures, notation and instrumentation can be developed through change processes.

For example, Marshmallow Laser Feast's *Forest*[31] was an installation made up of multiple poles, each with a note. Audience members moved through the space and hit each pole, generating the corresponding synthesised notes. The audience were responsible for the rhythm within the work while the artist had control over the pitch. The work was successful as the notes chosen combined to make the harmonic material. This removed the challenge of audience members entering and leaving the work. Similar to the loops and layers methods, this approach of having a set range of notes does lead to an installation that must be developed using one key (and in this case one chord) and can risk the work feeling quite static. This in itself is not an issue and, if intentional, can feel very meditative in quality. The form of installation used in *Forest* relies on the audience for an inciting interaction to start the sound and so requires intuitive forms of interaction or the potential for accidental interaction, e.g. in *Forest* the poles invited you to push them, but were also placed close enough together than an audience member may accidentally walk into one of them, triggering the interaction.

Similarly, the game *Ape Out*[32] uses a generative score reliant on how the player approaches the game. The percussion generated throughout provides a jazz soundtrack but rewards particular gameplay, you can play the game in a stealth mode and have a calmer soundtrack experience or play more violently for a more explosive sound.

Another potential option is to build the sound from the audience using the sound and interaction in the space to trigger processes, e.g. through using found sound made from audience speech or audience movement to trigger a process. Again, this provides an instant interaction and a composition process that builds on audience actions and interactions that rely on the audience behaviours and the audience understanding the interaction that is expected from them, providing an interesting risk for the work as generative processes may never have the source material to create a satisfying sound for the interaction.

All of these generative or unstructured processes provide an interesting challenge for the composer. Giving up so much control over the sound and composition means that the composer may never hear the outcomes of their composition; multiple versions of the work will be written and rewritten throughout the lifetime of the piece, heard by the audience but never by the creator.

Conclusion

Composition and sound design practices change in an interactive audience due to the roles of the audiences and the structural requirement of the form and the narrative.

In a sound design context, considering all sound within the categories of immerse and inform allows the sound designer to develop their approach with audience expectations in mind. This supports the practical development of audio as an interface using Nielsen's Usability Heuristics for Interface Design,[33] built using an understanding of semiotics, and supports the sound designer to consider the narrative immersion of the audience.

Within composition the composer is working with structures that keep the phase space open, providing multiple options within the work. Here a choice needs to be made between the amount of structure that is required in the work, which is determined by the narrative functions of the work, and where the work is placed. The different styles of composition in interactive music build on techniques developed through harmony and counterpoint and experimental 20th century music ranging from minimalism, use of moment form and more indeterminate techniques. Through this approach the composer is able to move further away from traditional linearity in music to provide more flexible approaches that can be adapted to a range of settings and uses.

The key technique needed across sound design and composition is the ability to think simultaneously in the micro and macro, considering the overall needs of the work, the finer details of the piece and the audience needs and experiences. Sound design and composition in an interactive space require the generation of vast quantities of material, or generative processes, that the composer and sound designer may never hear. This style of working requires a composer to be empathetic to audience needs and to adopt different audience/player stances to test the work.

Reading group questions

Within this discussion participants may wish to reflect on their own creative practices or use this as a tool to analyse the interactive work of other composers.

1   Using an example of an interactive piece of work and the framework of immerse and inform, what functions are different sounds serving within the work? How does the audience form the interface?
2   How can the composer and sound designer make sure that they are truly approaching their quality tests from different audience member's or player's perspectives? How could other techniques support this?
3   What is the impact of semiotic associations being different for different countries or territories? What does this mean for installations that might tour or for games being released in different territories?
4   How can approaches to build realistic immersive soundscapes be adapted to non-real world environments?
5   With the amount of material being generated, some of which can never be heard by the composer, how can they be sure of the construction/effectiveness of the work?
6   Using an example of an interactive piece of work, can loops and layers be identified in the construction of the music? How have these been used?
7   When structuring interactive composition using a loops and layers approach, what approaches could be used where key and time signature changes are implemented effectively within the looping sections?
8   While moment form provides a potential structure, is it possible to really create a work where each slice of the music is of equal importance and contains its own structure? How can this be achieved?
9   Working with more generative forms, how can the composer be considered the owner of the produced work?
10 What other creative practices include a focus on both micro and macro details? How can these be used as a comparison to support the development of interactive sound and music?

Notes

1 These are referred to by Collins using the term sonic envelopment, defined as 'the sensation of being surrounded by sound or the feeling of being inside a physical space (enveloped by that sound)'.
    Karen Collins, *Playing with Sound: A Theory of Interacting with Sound and Music in Video Games*. (Cambridge: MIT Press, 2013), p. 54.
2 *FIFA 1998* (EA Sports, 1998).
3 Jakob Nielsen, '10 Usability Heuristics for User Interface Design', *Nielsen Norman Group* <https://www.nngroup.com/articles/ten-usability-heuristics/> [accessed 12 March 2024].

4 I have written about how this can apply to teaching interactive sound and music in a chapter in *Teaching Electronic Music*. This section represents an extension of this prior writing: Lucy Ann Harrison, 'Teaching Principles of Interactive Sound: A Practice-Based Approach', *Teaching Electronic Music*, 2021, 90–102 <https://doi.org/10.4324/9780367815349-7>.

5 Collins outlines this in her writing on synchresis and audio as a learning tool in games: Collins, pp. 19–38.

6 The question of how interactive the work is has been addressed in chapter 1.

7 Christopher Small, *Musicking: The Meanings of Performance and Listening* (Wesleyan University Press, 1998), p. 39.

8 Nick Yee, 'Gaming Motivations Group Into 3 High-Level Clusters', *Quantic Foundry*, 2015 <https://quanticfoundry.com/2015/12/21/map-of-gaming-motivations/> [accessed 12 March 2024].

9 Michael Billington, 'The Duchess of Malfi', *The Guardian*, 14 July 2010 <https://www.theguardian.com/stage/2010/jul/14/duchess-of-malfi-review> [accessed 10 March 2024].

10 Sarah Atkinson and Helen W. Kennedy, 'From Conflict to Revolution: The Secret Aesthetic, Narrative Spatialisation and Audience Experience in Immersive Cinema Design', *Participations*, 13.1 (2016), 252–79.

11 *Batman: Arkham Asylum* (Rocksteady Studios, 2009).

12 Collins, p. 5.

13 *Batman: Arkham Asylum*.

14 Designs of pedestrian crossings are an interesting example of this. In some countries a light is shown to indicate that the button has been pressed while in others a countdown sound is played.

15 This may particularly affect the enjoyment of audience members or players with a more exploratory approach to a space.

16 This will be discussed further in chapter 6 on musical work and archiving processes.

17 This links to audience expectations as outlined in chapter 2.

18 Philip Pullman, *Dæmon Voices: Essays on Storytelling* (Oxford: David Fickling Books, 2017), pp. 20–22.

19 Pullman, p. 20.

20 This is different to audio in games that do not prioritise narrative such as match three games like *Candy Crush* or sofa co-op games such as *Overcooked*.

21 Billington, 'The Duchess of Malfi'.

22 Although, realistically, this may happen in some households particularly where games are shared.

23 See player motivations in chapter 2.

24 *Hades* (Supergiant Games, 2018) <https://www.supergiantgames.com/games/hades/>.

25 Middleware such as Fmod and WWise have functions that facilitate this.

26 Again, middleware is designed to facilitate this function.

27 Riley, Terry, *In C*, 1966.

28 *Star Wars Jedi: Fallen Order* (Electronic Arts, 2019).

29 Karlheinz Stockhausen, *Momente*, 1961–64.

30 I usually expect audience members to stay in each section of an installation for approx. 8–10 minutes and will usually make the soundscapes for my installation approximately 20 minutes long.

31 Marshmallow Laser Feast, *Forest*, 2014.

32 Gabe Cuzzillo, *Ape Out*, 2019.

33 Nielsen.

References

Atkinson, Sarah and Helen W. Kennedy, 'From Conflict to Revolution: The Secret Aesthetic, Narrative Spatialisation and Audience Experience in Immersive Cinema Design', *Participations*, 13. 1 (2016), 252–279.

Billington, Michael, 'The Duchess of Malfi', *The Guardian*, 14 July2010 <https://www.theguardian.com/stage/2010/jul/14/duchess-of-malfi-review> [accessed 10 March 2024].

Collins, Karen, *Playing with Sound: A Theory of Interacting with Sound and Music in Video Games* (Cambridge: MIT Press, 2013).

Harrison, Lucy Ann, 'Teaching Principles of Interactive Sound: A Practice-Based Approach' in *Teaching Electronic Music*, 2021, 90–102, doi:10.4324/9780367815349-7.

Nielsen, Jakob '10 Usability Heuristics for User Interface Design', Nielsen Norman Group <https://www.nngroup.com/articles/ten-usability-heuristics/> [accessed 12 March 2024].

Pullman, Philip, *Dæmon Voices: Essays on Storytelling* (Oxford: David Fickling Books, 2017).

Small, Christopher, *Musicking: The Meanings of Performance and Listening* (Wesleyan University Press, 1998).

Yee, Nick, 'Gaming Motivations Group Into 3 High-Level Clusters', *Quantic Foundry*, 2015 <https://quanticfoundry.com/2015/12/21/map-of-gaming-motivations/> [accessed 12 March 2024].

Media and art examples

Batman: Arkham Asylum (Rocksteady Studios, 2009).

FIFA 1998 (EA Sports, 1998).

Cuzzillo, Gabe, *Ape Out* (2019).

Hades (Supergiant Games, 2018).

Marshmallow Laser Feast, Forest, 2014.

Riley, Terry, In C, 1966.

Star Wars Jedi: Fallen Order (Electronic Arts, 2019).

Stockhausen, Karlheinz, Momente, 1961, solo soprano, 4 choir groups, 4 trumpets, 4 trombones, 2 electric organs and percussion (3 players).

The Legend of Zelda: Breath of the Wild (Nintendo, 2017).

Chapter 4

How can you make interaction a meaningful part of the work?

When a composer works in an interactive medium, they may also be responsible for the interaction design, embedding this in the compositional work and ensuring that it matches the requirements of the music being written. This requires many composers to build cross-discipline skills or to develop collaborative working partnerships as demonstrated in organisations such as CCRMA and IRCAM.

As discussed in chapter 1, unless for an artistic need, a composer should not be intending to deceive the audience with level of interaction in the work; interaction should be a key part of the composer's aims for the composition. The level of interaction develops along a scale from perceived interaction to the audiences becoming active collaborators within the work. This scale balances with the artistic and narrative aims of a piece where a composer may choose to limit the interactivity within a work.

Additionally, as outlined in chapter 2, audiences approach spaces with their own expectations where they read into all uses of sound within a space to understand expectations of interaction. In these cases, audio acts as an interface to guide an audience around a space and support their interaction with the work. Composers can work practically with this understanding, building in signs and symbolism in the work that further the audience's interaction with this space, as outlined in chapter 3.

With the composer's intention and the role of the audience as collaborators in mind, a key challenge when creating interactive sound and music is to build meaningful interaction into the work. For interaction to be characterised as meaningful it needs to be linked to the musical intention of the work and to the audience's experience of the work. The interaction needs to impact the sound within the music and be controlled by the audience to change structures and outputs in the work. The work cannot exist without the interaction.

Having defined in previous chapters the importance to the artistic intentions of the work that the audio is interactive, alongside how audiences approach the space, this chapter builds on those approaches and looks at how interaction can become integral to the work itself. Focusing on public art and sound installations as a starting point, combined with reflections on my compositional

DOI: 10.4324/9781003344148-5

practice, this chapter looks at how interaction can be embedded fully into the musical work in order to build a meaningful experience for the audience and the artist.

Building interaction from the beginning

In order for interaction to be meaningful to the work it must be built in from the beginning of the compositional process, at the same time as developing the music and sound design.

While interaction can be retrofitted into a work, this requires taking a fully realised piece of music and deconstructing it into constituent parts to recontextualise the music. This act of recontextualisation can be used as an act of alienation, in the Brechtian sense of 'making strange', which highlights and draws attention to different areas of work that may not be as audible in a traditional performance process. This can be a useful tool for outreach or teaching projects that aim to show the construction of music, for example the Philharmonia's *iOrchestra* project which ran from 2014–2015[1] created an immersive and interactive space for people to explore an orchestra from the perspective of different instrument's seats, allowing people to explore how orchestra sounds are layered.

Outside of pedagogy and outreach framings, if we consider the meaningful development of interaction within a piece there needs to be a symbiotic relationship between the sound and the interaction where neither can exist without each other. The aim of the interaction is to realise the artistic intentions of the work and provide practical solutions to audience experience. This particularly relates to the following two points when considering the concept of authenticity in interactive sound and music as outlined in chapter 1:

- The composer intends for interaction to be fully embedded in the work, it is not an add on and it does not mislead the audience about the amount of interaction
- The interaction truly impacts the work, having a long-term lasting effect on the piece that can be seen by the audience through their actions[2]

Building this authenticity into the interaction requires the composer to consider how the audience will experience the work from the beginning. This can be supported through the planning stage of the composition. For example, practically speaking, within my compositional work I choose to build an outline of guiding principles from the beginning to act as a guide for the composition. One of the challenges in all composition is the amount of choice available to the composer. This multiplies when working in an interactive format as it includes multiple iterations of material through the compositional phase space, interaction design, audience collaboration and non-linear design. By placing limits on the work and planning the boundaries for the composition I am able to use an economy of means approach to provide a more creative space for compositional work. In order to make decisions on the interaction design and compositional choices I make opening decisions on the following areas:

- The type of the space being used for the piece. Often this can be an initial decision that impacts all other parts of the work as, if working in a distinctive space, it may suggest narrative or sonic elements
- The theme of the work
- The intended audience and their experience
- How I wish the audience to feel during the work, this can be a particularly important decision as it impacts the style of interaction and the sonic choices for the piece

With these areas defined, interactive and compositional choices can be made to support and enhance the work. The theme and intended audience will dictate the types of interaction, gestures used and other aspects of the interaction design.

Building in interaction design at the beginning of the development process also further supports the compositional process outlined in chapter 3. This provides the parameters for the composer to choose the most appropriate compositional techniques and begin to think in the way that the structure and interaction within the work requires. For example, if they are working on a narrative piece where the audience can choose to approach the narrative with different levels of intensity, they can build in from the beginning the levels of orchestration to provide that flexibility for a loops and layers approach, writing in compositional blocks rather than writing a linear piece and then returning to orchestrate. If they choose a more installation-style work they can begin to think about the impacts of structure and how long audiences will stay in the work.

It can be tempting to take a technology first approach to interaction design, especially if a new piece of technology has just been released. This makes a composer's work seem cutting edge and provides new opportunities to the audiences that they may not have seen in other locations. However, without the larger planning in place, work that centres the technology first before the larger requirements of the piece can seem like a tech demonstration rather than a cohesive piece of music. The interaction is leading the music and the audience experience, rather than working symbiotically with the music to create the audience experience. If you consider the central approaches of IRCAM and CCRMA, the technology and music develop at the same time, allowing for technological innovation to be considered within the same frame as musical innovation. One doesn't lead the other but both work together for the benefit of the work. As Pierre Boulez, who was a true advocate of technology in music, says: 'although virtuosity or technology can be inspirational, the composer's gesture runs the risk of being destroyed or diminished when they are given priority'.[3]

Ensuring the interaction is musically necessary

While we have discussed the practicalities of composing for interactive sound and music in chapter 3, there are some considerations that need to be taken into account to make the interaction a musically meaningful experience. For the composer, this boils down to the question of 'why does this work *need* to be interactive?'. While there is an aspect of novelty to creating interactive sound and music, this alone is not enough to sustain a piece for meaningful interaction. The question of 'need' is different to asking 'why do I want this to be an interactive piece?'. 'Need' suggests that the work can only exist in this form. The interaction is essential to the music, and the composer cannot realise their artistic intentions without including interaction.

The first reason why a composition should be interactive is a very practical consideration. This is where the interaction provides support for the narrative or for structures within a game or interactive space. For example, game audio needs to have adaptive or interactive elements to support the needs of a game. The medium requires added flexibility within the work that can only be achieved through interactive composition processes.

Harder to define is the need for interaction within an installation or experimental piece of music. The choice for interaction is an artistic decision for the music, rather than a practical choice. Here the composer must justify what the interaction brings to the piece that will not be achievable without this element. This is where the role of the choice of the instigator for the interaction or the audience becomes important to the work.

In some cases, this artistic need can be a political choice. For example, when engaging with communities through public art, interaction can become a tool to support engagement, ensuring that communities feel ownership of the work. They are able to collaborate with the artist and build their communities into the work, rather than having work imposed on them by someone outside of the community. Presenting a work with interaction embedded can also provide a framing for an audience to question political structures of elements. For example, in a game like *Papers, Please* the limited interaction and choices in the game highlight to a player concepts of complicity and safety in oppressive political structures.[4] In other cases, interaction can be a symbolic choice for example, in a piece working with an robot like Yuan and Yu's *Can't Help Myself*[5] the robot is anthropomorphised through the interaction, symbolising a powerlessness to control the outcomes of the work which an audience can empathise with through the robot.

The choice to include interaction can also be linked to the composer's intended audience and how they want communities to engage with their work within a musicking context.[6] Writing music for a concert setting imposes a certain structure on how audiences will expect to engage with the music. The composer is inheriting the concert hall approaches and behaviours established by previous generations and practices. While a composer can aim for a more relaxed environment that reflects their ethos, this will not necessarily be affirmed by the venue and performance practices. Placing the work in an interactive setting allows the composer to ensure that different rituals are applied to the work.[7]

For a composer the 'need' for interaction can be an artistic choice related to the ability to realise their work outside of conventional structures, building on the experimental practices by composers such as Cage, Stockhausen and Cardew. These composers sought to expand the musical languages and possibilities of their composition through forms such as text pieces, graphic scores and aleatoric composition. Linear composition can be a very restrictive form. It requires time bound approaches and, if performed live, scores that can be played. Interactive work allows composers to play with the temporality of their work, building on a scale and opening up possibilities that they would normally close down. Interactivity provides the chance to work creatively across the phase space, realising more of their musical ideas in an open and playful way. The way interactive audience works collaboratively with audiences gives the composer the possibility to work directly with their audience rather than through a mediator. These considerations when creating interactive sound and music justify a compositional 'need'.

The audience as collaborators

As outlined in the definition of authenticity for interactive music in chapter 1, 'the audience or instigator of the interaction should be central to the work as a collaborator and the work should have been designed with them in mind from the beginning'.[8] An element of meaningful design in interactive sound and music is ensuring that the audience is able to recognise that their interaction is having an impact on the work being created. As previously mentioned, audiences can begin to test the limit of a work if they believe that their interaction is not having any impact on the work being created. This can include repeating gestures until they notice a result within the work.

Interaction should be instant and visible to the audience, ensuring that they are able to see the direct effect of their actions. For example, in the Checks web app produced as part of IRCAM's CoSiMa research into using mobile devices for collaborative interactive projects, the sound reacts instantly to the user's phone movement.[9] There is no delay between the gesture made and the sound produced. Within this app, the sound is continuous and adapts with the movement on the phone so that the audience can build an understanding of how they are influencing the sound created. The Checks web app was built to sound check the technology prior to a performance, ensuring that audiences were able to see how they were impacting the sound and demonstrating how gesture interaction can be used en masse before building to more complex musical interactions.[10] In a more physical demonstration of instant feedback, Anders Lind's *Voice Harvester* was an exhibition used to generate material for a larger work.[11] Lind needed to collect a range of voice recordings and needed to provide an engaging interface to encourage audiences to interact with the work. In *Voice Harvester* users speak directly into a recording device. The sound is recorded and their voices are manipulated in real time. Vibration from the speaker is used to agitate various materials including glitter and liquids, showing a playful, physical approach which allows the audiences to see how their voices are being recorded and encouraging more engagement with the piece.[12]

In this way, both these examples demonstrate a platonic ideal of how to show instant impact on interaction, but they are both abstract, standalone works with a practical function for larger pieces. They do not work within wider musical application of instant and impactful use of the audience as collaborators.

In a musical context there are some practical challenges and some musical considerations when creating instant interaction. For example, if you want to work with echoes based on audience voices and live sound input, in an interactive space you are not able to put the usual monitoring considerations in place that you would have on stage. This means that, if you are aiming for truly live audio, you will get instant and continuous feedback (that can also be quite painful for the audience). This could potentially be used musically in the style of tape loop music, sculpting the feedback in a similar way to Alvin Lucier's work,[13] but if the composer is aiming to continuously add to the voices in a space this will be limited through the feedback. While counterintuitive, it can be beneficial to record sounds into a small buffer before processing and replaying into the space. This reduces the need for the mic to be continuously open and allows for audiences to hear the almost instant feedback of their own voices without compromising the overall sound.

To really support the audience role as collaborators central to the interaction, part of the work needs to be left almost 'incomplete' by the composer. There has to be a space for the audience to contribute to the work. This can be thought of like a jigsaw puzzle where the audience is selecting material from the compositional phase space to fit into the work. These sounds can be chosen at random from pre-recorded options, an active choice by the audience who can see the musical options (like a 'choose your own adventure' through the music) or they can be triggered by other choices that an audience can make to follow a path or approach the world in a way that demonstrates their player motivations. For example, in a AAA game such as *Star Wars Jedi: Fallen Order*, the player can see the instant impact of their choices in the world through the sound cues that are generated and the level of intensity of music that is generated.[14]

Alternatively, the composer can leave building blocks of materials for the audience, where they can see the instant impact of how the work is assembled in the piece, giving more creative control over the finished piece. This can be seen through interfaces such as the Reactable[15] or Roli's NOISE app,[16] where each block can be joined and adjusted to create a new piece of musical work. Here the audience or user is almost assuming the role of the composer by being guided to musical choices such as time signature and key signature.

With all these options for building interaction with the audience as colla- borators, we again reach the central question of how much the composer should control the work and the output of the interaction. While free interac- tion as shown by the Reactable or NOISE app are fun outlets for the audience to use, they can sometimes lack musical structure and the composer's intention for the work can be lost in the piece. For the audience to truly feel like colla- borators in a work, rather than testers of technology, there must be a feeling that the interaction is building towards a greater musical aim. This is where the composer's role is to provide a clear structure and parameters to the interac- tion, ensuring that the instant response required by the audience is balanced with the greater musical and artistic needs of the work. For the work to be considered meaningful, therefore, counterintuitively the composer may need to limit some of the interaction.

Building on audio as an interface

When developing audio as an interface, there are musical and audio decisions that must be made using semiotic associations, as outlined in chapter 3, to ensure that the audience understand what is expected of them within a space. The audio acting in an interface role should also be immediately obvious to the audience to support error prevention within the work. Sounds should build on existing symbolic associations that the audience may be aware of from other cultural and musical settings to support them within their decisions.

If we are considering the roles of audio within the categories of sound to immerse or to inform,[17] the composer must make a decision within their crea- tion of the audio interface about how obvious they wish this instructional sound to be within the work. Do they wish for the audience to perceive that they are working in a virtual space with a fixed interface, or do they wish the audience to feel that their actions are more instinctual?

If the composer decides to prioritise the informative properties of the audio in an interface role, the mix of the audio will sit on top of the more immersive elements in the work. For example, in a traditional platformer game like *Sonic the Hedgehog*[18] the instructional audio for elements such as ring collection, jumps and injury sit in a soundscape above the musical soundtrack. This makes the interactive elements immediately obvious to the player, supporting their understanding of interaction in the work and allowing the audience to select the audio choices in the same way that they would understand visual interfaces that appear in a layer above the game. This can extend to support solving puzzles, where fragments of melody can be used to give clues to object placements or repeating patterns similar to the game *Simon*.[19]

If a composer chooses to emphasise immersion, then they may choose to embed the interface narratively within the work, using naturalistic sound to subtly guide an audience around a space. This can include the placement of voices and sparkly sounds in key locations to support audience movements and can also include symbolic associations that guide audiences towards an action. For example, in an escape room or games environment the sound designer may wish to draw an audience's attention to a talisman object that will support them using diegetic sound. For example, if they wish to draw attention to a radio, they may have the music change often and have the sound interspersed by static.

The level of immersion that a composer is aiming for through the sound and the role of audio as interface will impact the approach that the audience takes to the space. If a composer aims to fully immerse the audience in the space with naturalistic interaction and diegetic sound, the audience may use more imaginative play, placing themselves in the role of the actor, while a more informative, less naturalistic interface may support more focused and goal-led interaction and exploration in the work.

Embodied interaction

In order for interaction to be effective and meaningful for the work, the audience needs to understand what is expected of them with minimal instruction for how to engage with the work. This reduces the time lag between entering a space and engaging with the work.

The most effective way to embed interaction for an audience is to look at approaches that build on their current practices and understanding to create an intuitive interface. If we look at Nielsen's heuristics for interface design, the principle that the interface should have a match between the system and the real world is key.[20] If an audience member is required to learn a range of new bespoke gestures to interact with a piece of work, they may feel inhibited by the unfamiliarity of the action or struggle to trigger interactions. It is for this reason that we can see people struggle with new phones. For example, the upwards swipe is a gesture that unlocks a phone. If this then changes via a software update to be a left or rightwards swipe, it slows down a user's understanding of the tech, causing frustration. Similarly, there are some very powerful instruments like Roli's Seaboard,[21] which appears to be like a normal keyboard but embeds additional slide, vibrato and pressure gestures. While intended to be intuitive, pianists and keyboard players with more traditional backgrounds will take some time to adapt to this different way of working.

By using recognised and existing gestures we can embed 'embodied knowledge' within approaches to interaction. Concepts of embodiment extend from the writing of Merleau-Ponty where he discusses differences between the perception and embodiment.[22] Embodied knowledge is knowledge that is held but has become instinctual through time and repeated use, such as riding a bike. Embodiment suggests that the audience is able to react in the moment without over-intellectualising the actions required for the music, allowing for them to further engage with the musical, artistic and narrative elements of the work.

Within interaction, we are aiming for embodied gestures; gestures that are so entrenched in an audience's every day that they are instinct. As with semiotic associations, which provide layers of symbolic meanings built on cultural and historical understandings of music, audiences have an understanding of actions or gestures that they have developed through social and cultural practices, for example swiping gestures on smart phone interfaces. These gestures become second nature to people and become part of their natural 'toolbox' of inter-active approaches when looking at a new interactive work or technology. Similarly, as professional skills build and develop some embodied knowledge is held within our professional approaches to technical work, for example under-standing effective speaker placement within a site-specific installation. This embodied knowledge is also referred to as our 'professional instincts'.

By using these embodied gestures when building interaction into a work the composer can remove the fear of the unfamiliar when interacting with a piece, reducing the audience inhibitions within a space. This will also reduce the learning curve needed to interact with a piece of music. This is why an instal-lation such as Daily tous les jours' 21 Swings is such an effective example of interaction design (and why it has been used more than once as an example within this book!).[23] The use of playground swings provides an intuitive inter-face that builds on the embodied knowledge and gestures that the audience have used in childhood. Audiences do not require an instruction for how to interact with the work, they can walk up to the space and know what is expected. Furthermore, the swings are an enticing invitation to the audience, asking them to join the space and encouraging play.

While some elements of embodied gestures will be consistent across genera-tions, e.g. using swings, others may change by generation. For example, while a universal gesture for a phone call was based on a landline and used a thumb and finger to represent the ear and mouth pieces of the technology, children now use a flat palm when representing a phone to better symbolise a smart-phone. This has implications for interaction design and engagement. For example, if a composer were wishing to design an interaction that involved answering a phone through providing a relevant gesture, they may find that different generations will struggle to trigger the relevant audio response.

An extension of embodied gestures that can be applied to support audience engagement with interaction is the use of symbolism or mime-based gestures that represent the action linked to the interaction or intended audio output. Gestures with symbolic associations use a more metaphorical approach to sup-port the use of imagination and fantasy within an interaction space. For example Chris Milk's The Treachery of Sanctuary uses the audience's shadows to engage metaphorically with themes of birth, death and transfiguration.[24] In one of the screens the audience transforms into a winged creature, the required interaction is indicated through the development of feathers along the audi-ence's arms. When the audience member then flaps their arms, in an imitation of a bird, the character then flies away. Through using the playfulness of these childhood gestures and mimes, the composer can build on the audience's exist-ing knowledge to encourage further interaction. This requires symbolism that is widely understood and closely linked to the required actions, building on semiotic understanding of gestures and action. For example, if a gesture was being used to make an ascending musical melody, a logical gesture would be one that moves upwards and to the right representing not only the rising sound of the music but the temporal quality represented through the linear motion.

Making the most of liminal spaces

As outlined in chapter 2, the audience's perception of space is key to establish an environment for them to move into more playful and exploratory behaviours. This transition between the real world and the interactive space is achieved through passing through a liminal space or boundary. The liminal space creates an antechamber for the interactive work and can be used to build anticipation or build understandings of the required interaction and approaches to be taken within the space.

In interactive sound and music in an immersive or gallery setting, liminal spaces start within the ticketing and entry space to the work, including queuing systems. These spaces act as a holding space or an antechamber, providing practical solutions when an exhibition has high audience numbers but also providing a way to build anticipation for an interactive work. These antechambers can be used to establish the conditions of the work through atmospheric sound and music but also through extra material, web pages and apps that can provide more establishing information for the work. If you consider an immersive museum such as the York Dungeons,[25] the queue provides an opportunity to establish the interactive nature of the space with actors speaking with people in the queue, establishing that there will be conversation with actors throughout the exhibition. This also allows audiences to make an informed decision about whether this is the kind of exhibition that they will enjoy, allowing them to exit the queue if they feel this level of interaction is not appropriate for them.

If narrative information is required when understanding the work, this backstory can be provided through antechamber spaces. Punchdrunk demonstrated this at *The Burnt City* through their queue spaces which included artwork of family trees and supporting information on character relationships, including pictures of the actors in character so that they could be recognised within the space.[26]

Liminal spaces can also be used to provide a break in interaction for audience members if they find that they are overwhelmed with experiences or if they would like to pause and return to this work. Again, Punchdrunk provided a liminal space within *Burnt City* by having a cabaret bar space, still linked to the narrative but with more relaxed rules. In this space audiences were allowed to speak and could remove their masks while they got a drink. This liminal space also provided additional backstory to characters who were silent within the space through the MC's narrative, dances and choices of songs to support audience understanding of the narrative when they returned to the space.

In a game environment these liminal spaces are provided through non-game actions such as menus, pause functions and loading screens. These spaces can be signified through audio techniques such as differentiating the music or providing dynamic filtering that triggers when the live menu is pressed. This can be applied to in-person interaction through providing clear musical cues that signify the liminal spaces, providing a sense of place but making it clear that the interactive processes have not yet started.

Liminal spaces can also build more gradually and be less structured. For example, when attending a football match the buildup and anticipation starts in the walk to the stadium where the spectators begin to move from individuals and small groups into a larger crowd, demonstrating Hill et al.'s model for building anticipation among crowds.[27] Here the liminal space is adaptive and temporary, forming before and after the event and disappearing after the crowds disperse. A sense of this gradual build can be created through the pre-experience materials being sent to audience members, establishing the needs of the work and requirements in advance. This is the structure used by Secret Cinema to build anticipation, where audience members are sent character information and background information for the event prior to attending.[28]

By working with liminal spaces, the composer is able to set conditions and understanding for interaction, preparing the audience and ensuring that they are able to engage successfully and meaningfully with the work.

Individualised audience experience vs audience swarms

As different audience members approach the space looking for individualised experiences while also looking to engage with wider communities, getting a sense of the 'aura' of others within the space, interaction aiming for larger group projects needs to support both of these approaches while ensuring that audiences have parity of experience. As we have discussed in chapter 2, audiences in interactive environments struggle with FOMO (fear of missing out) which causes them to approach an interactive space with a collector's mindset, trying to find as many experiences as possible. This approach can lead to audience swarming and also create a pretty unsatisfying audience experience. If you are an audience member participating in a swarm there will always be the feeling that you have missed something special and need to move to the next event or experience in a work. Similarly, this bunching of audience members can mean that people not near the centre of the work cannot see or interact with the experience, leading to a reduced experience of the piece. While staggered audience entry times can support interactive events to ensure that the audience experience is preserved, there are techniques in interaction design that can support larger audiences.

The most challenging option to ensure meaningful interaction on a large scale is to build an interactive interface that works with crowds, where the work can build with larger audience numbers without reaching a saturation point of interaction. For example, Umbrellium's work *Marling*[29] is built with mass participation in mind. The piece uses lasers and microphones to provide a live visualisation of the waveforms of the audience's voices. The larger the audience, the more visuals that they can see within the work so as the audience grows so does the piece. Audiences do not need to work as a collective with strangers to achieve this outcome, it happens naturally with all audience members acting as individuals within the piece. This style of large-scale interactive design is hard to achieve as it involves visualising the work and the outcomes on a large scale. A large, simple interaction is more achievable on this scale than smaller detailed options. The ability to provide instruction to individual audience members is reduced, and when working in a large group audiences are more likely to follow the example of other audience members than read instructions, as large-scale works such as this will be likely to cause a honeypot effect with audience members being drawn over to the space having seen others engage publicly.[30]

Simplicity in audio is also important for this scale of interactive work as smaller soundscapes will begin to sound cluttered when layered across the scale of audience members collaborating with the work. For example, Daan Roosegaarde's *Dune* consists of a series of bushes that whisper as the audience gets close to the piece.[31] The simplicity of this sound is it easily multiplies and extends across a space. The higher frequencies mean that it is a locational sound so that the whispers are placed where a person is standing, giving other audience members the sense of where other people are in the space. The work is also modular so can be extended for larger spaces and events.

Another option is to consider the number of potential audience members when designing the interaction for the installation so that this can be incorporated into the initial design to create smaller points of interaction within the work, allowing multiple audience members to interact with different parts of the work simultaneously. If a point of interaction is busy, the audience can then move on to a different part of the space. This is similar to playground design; if the slide is busy a person can move on to the roundabout or the swings. This technique of creating multiple interaction points is used in *Assemblance* by Umbrellium,[32] an installation where audience members can interact with different 3D light structures to produce visual and sonic results. Some of these structures work individually while others can be explored collaboratively with other audience members. All the pieces run concurrently so audiences are provided with options, reducing bottlenecks in the work.

With more narrative or structured pieces of work, audiences will be looking to piece together the overall piece through their searching for interactions. If there are live performance sections or elements of music that need to be heard by the audience, as with Punchdrunk's *The Duchess of Malfi*,[33] the composer can play pieces on staggered cycles, treating each moment of interaction as a separate repeating loop. By repeating each cycle multiple times, the audience do not feel under pressure to chase the different events, allowing them more time for free exploration.

While these experiences support large group engagement in a work, there is still the power to add 'easter eggs' in an interactive piece. These surprises allow for audiences to find smaller individual experiences, rewarding their exploration. These hidden rewards in a piece become shareable experiences for the audience, acting as a promotional tool for the work. For example, in the 2018 theatre piece *Memories of Fiction: The Living Library*,[34] the work started with an interactive exploratory space. These consisted of listening posts and a touch sensitive interactive bookcase that told people's experiences of libraries. Additional audio was hidden inside books using a speaker system that started playing as the audience member began to read a book, acting as an additional experience for people who chose to look further into the books in the set.

Conclusion

Building meaningful interaction into a piece of interactive sound and music relies on the music and interaction being developed simultaneously. The composer needs to have a strong sense of the need for the interaction, aligned with their artistic intentions and the practicalities of the work.

The central question of why interaction is taking place in the work is the most important to understand how meaningful this interaction is. This gives a piece purpose and focus, allowing for directed interaction that fully links to the musical needs of the work. For a composer working in an interactive space this choice is a 'need' related to musical aims that sit separate to linear or fixed constraints of composition in a performance context.

For the interaction in the work to be meaningful the composer is required to consider their collaborator within the work as the audience or the instigator for the interaction. It is this collaborator who will be responsible for working with the interactive element and they should sit at the centre of meaningful interactive design. Therefore, the music within the work should leave space for the collaborator to either choose musical solutions to the work, build musical structures or build the work from tools provided by the composer. The collaborators in the work are supported by the audio interface created by the composer where the roles of immerse and inform support the decision of how obvious the audio interface is within the work. This links to choices about how immersed the audience is supposed to be within the space; if they are intended to use imaginative play within a space a more embedded and subtle audio interface will support the narrative immersion.

To further support audience interaction the composer can make use of embodied gestures, or gestures with symbolic associations that the audience use within their everyday experiences so that the interaction is instinctive to the audience. These gestures can link to the musical work being created, allowing the audience to work in a similar way to physically sculpting or painting the music. Liminal spaces can set a space to prepare audiences for interaction and to ensure that they understand the requirements within the work, and also provide spaces for interactive work to be ethical in how audiences are treated by allowing audiences to choose whether they wish to engage with the style of interaction and providing spaces for audiences to take a break from the interaction. When supporting interaction in larger groups the composer can look either at large-scale, simple interactive structures that support mass engagement or at smaller interactive experiences, and loops that run concurrently giving the audience multiple choices for interaction while preserving their audio experience.

How meaningful interaction is within a musical piece really lies in the intentionality of the composer's choices. This interaction sits beyond the novelty of interaction and new technology in creating a true collaboration between the composer and the audience.

Reading group questions

1   How measurable is 'meaningful' as a term for interaction? Can this truly be defined?
2   Can interaction be retrofitted onto a project outside of an outreach or education context and still be meaningful? How can this be achieved?
3   How can the composer's intentions be shown through their choice of interaction?
4   What other compositional approaches would work for a composer wishing to realise their work in a non-linear format?
5   What is an example of an instinctive interaction? How can this be developed?
6   How are embodied gestures changing across generations? How can these changes be used when creating meaningful interaction in a piece?
7   How can audio as an interface be used while still contributing to narrative aims of the work?
8   How can a liminal space be developed digitally in order to build anticipation for an interactive piece of work?
9   How can a complex piece of interactive music function on a large scale?
10   How can cohesion be created in multiple smaller pieces of interaction in a large-scale work?

Notes

1   Philharmonia, *iOrchestra* (South West England, 2014).
2   See chapter 1.
3   Pierre Boulez, *Music Lessons: The Collège de France Lectures*, ed. by Jonathan Dunsby, Jonathan Goldman, and Arnold Whittall, trans. by Jonathan Dunsby, Jonathan Goldman, and Arnold Whittall (London, England: Faber & Faber, 2018), p. 167.
4   Lucas Pope, *Papers, Please*, 2013.
5   Sun Yuan and Peng Yu, *Can't Help Myself* (Guggenheim Museum, 2016).
6   Christopher Small, *Musicking: The Meanings of Performance and Listening* (Wesleyan University Press, 1998).
7   This, incidentally, was my initial motivation as a composer for moving to an interactive setting.

8 See chapter 1.

9 CoSiMa, 'CoSiMa | Collective Sound Checks' <https://ircam-cosima.github.io/cosima-checks/public/> [accessed 9 April 2024].

10 CoSiMa, 'Collective Sound Check @ Paris Face Cachée – CoSiMa', 2015 <https://cosima.ircam.fr/2015/02/07/experimentations-sonores-paris-face-cachee/> [accessed 9 April 2024].

11 Anders Lind, *Voices of Umeå: The Voice Harvester* (Umeå, Umeå University, 2014).

12 Daniel Fallman, 'The Voice Harvester' <https://www.dfallman.com/voice-harvester> [accessed 10 April 2024].

13 Alvin Lucier, *I Am Sitting in a Room*, 1969.

14 *Star Wars Jedi: Fallen Order* (Electronic Arts, 2019).

15 Reactable, 'Welcome', *Reactable Legacy* <https://reactable.com/> [accessed 26 March 2024].

16 ROLI Ltd., 'NOISE on the App Store' <https://apps.apple.com/gb/app/noise/id1011132019> [accessed 9 April 2024].

17 See chapter 3.

18 Sonic Team, *Sonic the Hedgehog* (Sega, 1991).

19 *Simon* (Milton Bradley, 1978).

20 Jakob Nielsen '10 Usability Heuristics for User Interface Design', *Nielsen Norman Group* <https://www.nngroup.com/articles/ten-usability-heuristics/> [accessed 12 March 2024].

21 ROLI Ltd., 'Seaboard RISE 2 | ROLI' <https://roli.com/> [accessed 9 April 2024].

22 Maurice Merleau-Ponty, *Phenomenology of Perception*, trans. by Donald A. Landes (London: Routledge, 2012).

23 Daily tous les jours, '21 Balançoires (21 Swings) | Daily Tous Les Jours' (Montreal, 2011) <https://www.dailytouslesjours.com/en/work/21-swings> [accessed 15 March 2024].

24 Chris Milk, *The Treachery of Sanctuary* (Barbican Centre, 2014).

25 Merlin Entertainments, 'The York Dungeon', *The York Dungeon* <https://www.thedungeons.com/york/> [accessed 9 April 2024].

26 Punchdrunk, *The Burnt City* (Woolwich, 2022).

27 Tim Hill, Robin Canniford and Giana M. Eckhardt, 'The Roar of the Crowd: How Interaction Ritual Chains Create Social Atmospheres', *Journal of Marketing*, 86.3 (2022), 121–39 <https://doi.org/10.1177/00222429211023355>.

28 Sarah Atkinson and Helen W. Kennedy, 'From Conflict to Revolution: The Secret Aesthetic, Narrative Spatialisation and Audience Experience in Immersive Cinema Design', *Participations*, 13.1 (2016), 252–79 (p. 267).

29 Umbrellium, *Marling* (Eindhoven, 2012) <https://umbrellium.co.uk/projects/marling/> [accessed 9 April 2024].

30 Wouters, Niels, John Downs, Mitchell Harrop, Travis Cox, Eduardo Oliveira, Sarah Webber, Frank Vetere, and Andrew Vande Moere. 'Uncovering the Honeypot Effect', delivered at 'Proceedings of the 2016 ACM Conference on Designing Interactive Systems - DIS '16', 2016. <https://doi.org/10.1145/2901790.2901796>.

31 Studio Roosegaarde, *Dune* (Lumiere, Durham, 2009) <https://www.studioroosegaarde.net/project/dune> [accessed 9 April 2024].

32 Umbrellium, *Assemblance* <https://umbrellium.co.uk/projects/assemblance/> [accessed 29 March 2024].

33 Punchdrunk, *The Duchess of Malfi* (London, 2010) <https://www.punchdrunk.com/work/the-duchess-of-malfi/> [accessed 9 April 2024].

34 Seadog Theatre and Laura Bridges, *Memories of Fiction: The Living Library* (Omnibus Theatre, Clapham, 2018).

References

Atkinson, Sarah, and Helen W. Kennedy, 'From Conflict to Revolution: The Secret Aesthetic, Narrative Spatialisation and Audience Experience in Immersive Cinema Design', *Participations*, 13. 1 (2016), 252–279.

Boulez, Pierre, *Music Lessons: The Collège de France Lectures*, ed. by Jonathan Dunsby, Jonathan Goldman and Arnold Whittall, trans. by Jonathan Dunsby, Jonathan Goldman and Arnold Whittall (London, England: Faber & Faber, 2018).

CoSiMa, 'Collective Sound Check @ Paris Face Cachée – CoSiMa', 2015 <https://cosima.ircam.fr/2015/02/07/experimentations-sonores-paris-face-cachee/> [accessed 9 April 2024].

CoSiMa, 'CoSiMa | Collective Sound Checks' <https://ircam-cosima.github.io/cosima-checks/public/> [accessed 9 April 2024].

Fallman, Daniel, 'The Voice Harvester' <https://www.dfallman.com/voice-harvester> [accessed 10 April 2024].

Hill, Tim, Robin Canniford, and Giana M. Eckhardt, 'The Roar of the Crowd: How Interaction Ritual Chains Create Social Atmospheres', *Journal of Marketing*, 86. 3 (2022), 121–139, doi:10.1177/00222429211023355.

Merleau-Ponty, Maurice, *Phenomenology of Perception*, trans. by Donald A. Landes (London: Routledge, 2012).

Nielsen, Jakob, '10 Usability Heuristics for User Interface Design', Nielsen Norman Group <https://www.nngroup.com/articles/ten-usability-heuristics/> [accessed 12 March 2024].

Reactable, 'Welcome', Reactable Legacy <https://reactable.com/> [accessed 26 March 2024].

ROLI Ltd., 'NOISE on the App Store' <https://apps.apple.com/gb/app/noise/id1011132019> [accessed 9 April 2024].

ROLI Ltd., 'Seaboard RISE 2 | ROLI' <https://roli.com/> [accessed 9 April 2024].

Small, Christopher, *Musicking: The Meanings of Performance and Listening* (Wesleyan University Press, 1998).

Wouters, Niels, John Downs, Mitchell Harrop, Travis Cox, Eduardo Oliveira, Sarah Webber, Frank Vetere and Andrew Vande Moere, '*Uncovering the Honeypot Effect*', delivered at 'Proceedings of the 2016 ACM Conference on Designing Interactive Systems - DIS '16', 2016. doi:10.1145/2901790.2901796.

Media and art examples

Daily tous les jours, *21 Balançoires (21 Swings)* | *Daily tous les jours* (Montreal, 2011) <https://www.dailytouslesjours.com/en/work/21-swings> [accessed 15 March 2024].

Lind, Anders, *Voices of Umeå: The Voice Harvester* (Umeå, Umeå University, 2014).

Lucier, Alvin, I Am Sitting in a Room, 1969.

Merlin Entertainments, 'The York Dungeon', The York Dungeon <https://www.thedungeons.com/york/> [accessed 9 April 2024].

Milk, Chris, *The Treachery of Sanctuary* (Barbican Centre, 2014).

Philharmonia, *iOrchestra* (South West England, 2014).

Pope, Lucas, *Papers, Please* (2013).

Punchdrunk, *The Burnt City* (Woolwich, 2022).

Punchdrunk, *The Duchess of Malfi* (London, 2010) <https://www.punchdrunk.com/work/the-duchess-of-malfi/> [accessed 9 April 2024].

Seadog Theatre and Laura Bridges, *Memories of Fiction: The Living Library* (Omnibus Theatre, Clapham, 2018).

Simon (Milton Bradley, 1978).

Sonic Team, *Sonic the Hedgehog* (Sega, 1991).

Star Wars Jedi: Fallen Order (Electronic Arts, 2019).

Studio Roosegaarde, *Dune* (Lumiere, Durham, 2009) <https://www.studioroosegaarde.net/project/dune> [accessed 9 April 2024].

Umbrellium, *Assemblance* <https://umbrellium.co.uk/projects/assemblance/> [accessed 29 March 2024].

Umbrellium, *Marling* (Eindhoven, 2012) <https://umbrellium.co.uk/projects/marling/> [accessed 9 April 2024].

Yuan, Sun and Peng Yu, *Can't Help Myself* (Guggenheim Museum, 2016).

Chapter 5

Is interactive sound and music ever finished?

There is a truism that art is never finished, only abandoned, which, from my experience, definitely applies to the composition process. As a composer you draft and redraft work but eventually the piece needs to be delivered or performed and can no longer be edited. However, there are composers that continue redrafts and revisions once the piece is out in the world and keep their pieces as a 'live' version that keeps adapting throughout their career. Stephen Davies outlines this process as having both artistic and creative concerns for the composer, as they work past the point of the work being considered 'complete' and released into the work.[1]

With interactive sound and music, even when the composer has delivered the work to the public completing their work on the piece, the work itself is still in state of flux. It is waiting for the audience in their collaborative work to complete the piece. This creates a condition where the composer never actually hears their finished work, or only hears versions of the piece. With these constantly changing versions of the music it is feasible that the work may never actually be considered finished.

This chapter will investigate the notion of completeness in interactive sound and music, looking at precedents from art and music. These concepts will be investigated both from the point of view of the composer and the audience as their collaborators. Through this, we will consider what the impact is on the composer if they can never hear the completed work. It will look at the implications on the composition process and the techniques that are used to review a piece before it is delivered to the public whilst considering whether interactive audio can ever be finished.

When to call time on the composition process

Composition is a finite process with an eventual aim to release a finished piece of music out into the world. While all composers take a different approach to their work, a unifying challenge for all composers is when to decide that the work is done and ready to be passed on to the listener or the audience. This is often related to the composer's instinct when creating the work. The distinction

DOI: 10.4324/9781003344148-6

between when a composer is pausing before redrafting a work or the composer has finished the piece represents the difference between internal and external processes, with only 'completed' work being available to external listeners. While still in the internal 'redraft' process, prior to being in a state where the composer feels comfortable with the work being heard by others, the work is still in a theoretical form for the composer with elements of structure, harmony and melody that still may be changeable. Once the work is in the external stage of completeness these elements of structure and compositional choices are more solidified and less open to being changed.

Stephen Davies outlines a number of conditions for indicating that a composition can be considered complete and in this external stage. These are:

- The composer's declaration
- The composer finalising the score
- The score being copied for performance
- Public performance of the work
- An authorised version of the score[2]

These definitions take an approach centred around the concept of the score that suggests the notation of the work is the sign that the work is completed.[3] Since interactive sound and music requires an open phase space of composition a completed score is unlikely to be produced for the work as this will not fully represent the context of the piece. Similarly, the performers and collaborators working with the piece, most often the audience, do so without requiring the prior preparation that an orchestra or similar performer would need prior to performance. Interactive sound and music goes straight from the composer completing the piece to the audience collaborating with the work live.

Within interactive sound and music, there are different signifiers of a work being completed that must exist separate to the concept of a score. For a work of interactive sound and music to be complete it is being released into the world within a game, gallery or a performance context. A composer's indication that the work is complete for interactive environments can be based on the following conditions:

- The composer's declaration
- The composer finalising the files for incorporation in the work – this will include loopable audio, stems and patches or files used for the implementation of the work
- Instructions for the assembly of the audio in a game or gallery
- Public performance of the work or release of the game
- Creation of artefacts of the work for promotion and further recreation of the work

For a composer to be satisfied that they are ready to meet these conditions they need to call time on the composition process. When working in a professional or commercial process, the decision may be out of the composer's hands, especially when working in games where there are tight turnarounds on release deadlines and galleries that may be working towards an opening night. The composer still needs to have processes in place to decide when they feel that the music is complete, but this will be balanced with the commercial needs of the work. If these commercial deadlines and practicalities are removed the composer is left with a bigger challenge on deciding the completeness of the work. In this case, the decision of completeness is an artistic choice.

Within the context of artwork Livingstone refers to 'genetic and aesthetic completion', where genetically complete work is considered complete when the composer or artist decides that it is ready to be released and a work is aesthetically complete when it has achieved its artistic aims.[4] This is an important distinction for interactive sound and music. The work being genetically complete is the point when the technical work that sets the conditions for audience interaction has been completed. This includes the pre-composed work, the audio that acts as an interface, the interaction design and any algorithms that are used to create generative sound and music. The aesthetic completion of the work is the point where it realises its artistic aims of collaborating with the audience at the point where the final piece of the puzzle for the work is completed. While the composer can decide when the work is ready for installation, e.g. the files and audio are ready to be implemented making the work genetically complete, the work cannot be considered aesthetically complete until the audience engages with the piece. It is through this interaction that the work meets its artistic and aesthetic aims.

Knowing whether a piece of music is genetically complete sits within the embodied knowledge held by the composer. This is part of the professional skills and instincts of the creator. Rohrbaugh refers to this knowledge of the artist as an asymmetry where the artist's view of the completeness of their work overrides outside critiques of the work, qualifying the condition of this with:

> If the fact of completeness just is a psychological fact about the artist, then the asymmetry we find here is just a special case of the fundamental epistemic asymmetry between knowledge of one's own mental states and those of others.[5]

As the creator, the composer can be seen as holding an authority over their own work. But this does still retain hierarchies of the composer's role where they are seen in an auteur role over the creative decisions of the music. In the case of completeness in the composition, this role is appropriate for the genetic completeness when setting the artistic conditions for the interactive work prior to the audience joining the piece.

From a practitioner's perspective this instinct to decide the completeness of the work is learnt and developed over time and is built on the following areas:

- **The structure of the work.** Similar to a sentence or story structure the music reaches a natural conclusion. In traditional classical writing this is when the main themes have been resolved. In experimental processes this is a more open structure but can be signified by the completion of the sonic experimentation or return to the original material
- **The sonic and timbral completeness of the work.** Similar to colouring a piece of visual art, is the work sufficiently coloured? This can be dictated through the limitations of the piece or can be a later decision on orchestration based on the emotional or technical needs of the piece
- **The realisation of the core idea.** Has the work completed the technical aim of the piece? For example, does the music support the needs of the media or has the work fully investigated the experimental aspects being explored in the piece?

When we teach composition, we support students to build this understanding of when their work is complete through their listening skills and through peer support and feedback. Playback sessions and peer review form a core teaching process that can then be extended into peer groups as students graduate into professional worlds. By observing how their peers understand if a work is finished, they can understand how to adapt and work with their piece. By providing clear deadlines and requirements for compositional exercises they can further develop their skills and artistic instincts so that they can adapt to professional needs. Eventually, this becomes the practitioner's instincts for their work. Once a composer has confidence in these skills through structuring and developing their work they can then begin to test and develop structures, shifting the perception of what sounds complete in a work.

The instincts required for the composer when determining completeness in interactive sound and music holds slightly different requirements than when working with a linear, fixed, piece of music. Interactive sound and music requires a change in approach, particularly with how the composer considers the structure of the work. With a sound installation or a piece that requires open exploration the structure is not looking to be completed, it needs to be kept open, therefore it is not aiming for a sentence or story-like structure where the key themes are resolved or the experimentation is considered complete. The notion of genetic completeness when related to interactive structures is further supported by and intertwined with the realisation of the core idea. This would provide these revised conditions to determine completeness in the composition.

- **The structure of the work.** Are loops and layers functional? Do generative composition algorithms function within the work? Will the structure fulfil the compositional aims if an audience member stays for varying periods of time?

- **The sonic and timbral completeness of the work.** In addition to the previous conditions of orchestration and timbre, does the audio functioning as interface for the audience function in the expected way and can this be used to support audience interaction?
- **The realisation of the core idea.** Has the work completed the technical aim of the piece? Are all aspects of the interaction fully functional and do they clearly relate to the music, and the artistic and functional aims of the piece?

The composer knows that the work is genetically complete when they feel that they have fully realised the interactive and artistic aims for the work. This includes ensuring that all interactive elements are working and the compositional building blocks are in place for the work to be realised once passed on to the audience. In a games context, this means that they have created music with flexibility to adapt to different locations and playing styles. Within experimental and sound installation contexts the composer can be satisfied that the work is complete when they know that it will be suit the space that it will be placed in and the format of the work is appropriate, allowing for the audience to stay for varying amounts of time and interact meaningfully with the work.

The composer can never hear their work in full

When determining if a piece of work can be considered aesthetically complete in an interactive context, the composer needs to consider how it will sound to the audience. While this would ordinarily include listening to the work in its entirety, this standard process for the composer would not be entirely possible in an interactive context. As outlined within the composition process in chapter 3, the composer generates a large amount of melodic and harmonic fragments to ensure that there are a range of musical options in place for the interaction within the work. This represents opening out the phase space in a composition process so all the composer's options are kept open throughout the piece. Playtesting functions can be used to determine the genetic completeness of the work by replicating audience behaviours including testing all the sound to the limits, testing loops and layers and playing through the work using the approach of different audience members and personas. This provides a practical framing to attempt to determine if the work is finished. However, the composer will never be able to test all aspects of the work to determine if it is genetically complete.

As this style of composition keeps the phase space open, every version of the compositional work has multiple other versions hidden inside it. Conditions of interaction will mean that some versions of the sound are randomly triggered while others are conditional on the action that happened before it. If a section of a piece had ten different melodic fragments that could be played in any order, with all being played once for the section to be complete, there would be 3,628,800 potential versions of that one section of music.[6] This number increases for every additional option put in and for potential repetitions of musical fragments, as would be usual in compositional practices. As more versions of the music are kept open in the phase space, the more challenging it becomes for the composer to truly test if a work is genetically complete through listening to the fragments. Instead, the composer must be satisfied that each fragment works in isolation and that each has been designed so that it can be successfully joined in any order with the other fragments.

Assessing genetic completeness becomes even more complex if a work is created using algorithmic or generative composition processes, for example a piece such as *Listen to Wikipedia*, which uses a live data stream.[7] Live generative composition exists in potential only prior to the instigator for the interaction being added to the work (this can be the audience interacting with the piece or the data stream being connected to the work). The composer can only set the conditions for the output and test with the data on a given day or data simulations. They must leave the rest of this process up to chance. In order to determine the genetic completeness for a generative piece of music, the composer needs to create a realistic test of the conditions in which the generation will happen, either through testing the piece themselves or through generating test data that accurately emulates the data the work will be using. This testing for generative composition using live data or audience interaction is an important process to consider readiness as live data can cause some bugs in a piece that haven't been identified during the creation process, for example, if the data refreshes continuously and in real time in the algorithm, it can create a strobing effect where the compositional process continuously starts from the beginning.

In interactive composition, considering if a work is genetically complete also includes conditions of the technical completeness of the work. The stability of an interactive platform is a key consideration when deciding if a work is ready to be released. For smaller pieces of interactive sound, and for game platforms, it is possible to test this readiness through play testing approaches. For large-scale interactive pieces it becomes harder to determine if the work will succeed at scale without platforms crashing. While a composer can run test models and aim to test processes digitally, this will still not test how all processes work when happening concurrently with hardware. The best approach that a composer can take in this case is to work with a test audience, although this is an expensive solution and will not fully be able to model audience numbers, particularly at a mass participation event.

The processes above provide approaches and conditions for testing the genetic completeness of the work and aim to provide the composer with an overall sense of how the work will function when it is released. While these only provide approaches for a small amount of the audio embedded in a piece, they give the composer a sense of how the work will sound on release and allow them to determine if it is genetically complete.

While the composer can put processes in place to test the practicalities of the work and determine the genetic completeness of a piece, they will only ever be partially able to hear the aesthetically complete work. While a work may be considered genetically complete, in a practical sense it is still waiting for an audience as a long-distance collaborator for the piece. In order to ensure that the interaction is meaningful it must rely on the audience as a collaborator, therefore a genetically complete piece of work is still musically and aesthetically incomplete. The work is waiting for the collaborator to enter the interactive space and finish the process to complete the work which will then give it the potential to be aesthetically complete. It is like Schroedinger's musical work: while the audience is still a potential collaborator the work is both complete and incomplete at the same time.

With game music, the composer is unlikely to ever see how the players interacts with their work. For longer form experimental or public art sound installations the work may stay in place for a prolonged period of time, weeks or months (potentially longer). This means that the audience are collaborators that the composer will never meet, like a silent remote partner in the work. As the audience are considered long distance collaborators, for the composer, this means that they will not necessarily see the outcomes of their collaboration. While they have a sense of their musical work, they will only see snippets of it fully realised. What they will see is the potential of what they have created but never the full outcome, or a few versions of the full outcome but not every single possible outcome.

For the composer, not being able to see their piece fully realised makes the outcome of their work more like a thought experiment than a solidified, completed piece of music. They have set the conditions of the work and solidified aspects of the structure and musicality but the work itself exists in the moment only with the audience, as part of that audience's experience. Through releasing their work out into the world, it goes from being a series of files and musical ideas to being something more ephemeral that exists briefly with each audience member before dispersing completely once the audience interaction is complete.

The composition existing to the composer as a thought experiment rather than as a solidified piece of work is similar to composers working within experimental indeterminate musical forms such as text pieces and graphic scores as demonstrated by the 20th century experimental composers such as Cage, Cardew and Stockhausen. Text pieces and indeterminate work rely on the performers to interpret the instructions in order to create the work. Notation is not usually provided, leaving compositional decisions to the performers as collaborators. The conditions for their work are set but they will never see the work fully realised. In the case of some experimental musical works, the potential for a work to be fully realised is not even possible, for example in Gold Dust from Stockhausen's *Aus Den Sieben Tagen* the composer requires the performer to live completely alone for four days without food, sleeping 'as little as necessary' and thinking 'as little as possible'.[8] This would require quite a lot of commitment from a performer and would not be possible to complete while maintaining a reasonable working environment.[9] These pieces act more as a theoretical thought experiment or a stimulus for a meditation that were not heard in every iteration by the composer and continue to be interpreted in different forms and approaches.

As the composer will not be able to hear all the iterations of their work, or the aesthetically complete piece, those writing music using forms such as those used commonly in interactive sound and music are required to have a finely tuned musical imagination. The composer needs to be able to view the potential of the work when constructing the pieces, imagining the possible audience member collaborating with the piece and how this will change the sound and structure of the music. Composers need to be able to imagine different iterations of the melodic and harmonic fragments and imagine these in non-linear functions. It is through this act of musical imagination that they are able to consider what an aesthetically complete piece of work will sound like before they release it to their collaborators.

The composer's acceptance of control and collaboration

As the audience are responsible for the aesthetic completion of a piece of interactive sound and music, their role as collaborators in the music becomes central to the question of if the work is complete. In discussing collaboration within art practices, Livingston and Archer discuss the agreements made by artists working collaboratively when deciding if a piece of work is complete. This requires all artists involved in the project to agree on the rules and procedures for the collaboration and their artistic aims. These requirements may change as the development of a piece progresses to reflect the changing values of the collaborators or the changing social dynamics between collaborators.[10] As previously discussed, interactive sound and music is a long distance collaboration between the composer and the audience, where the two collaborators never meet or speak and only communicate through the conditions of the work. The rules and the condition for the collaboration are established by the composer in the creation of the work through the way that audio acts as an interface for the audience, the type of interaction in the piece and any extra materials that are given to the audience to support their understanding of the requirements of the work, e.g. instructional plaques or a programme for the event. Unlike with a collaboration between two artists, the rules are not always mutually agreed between the parties, the audience as collaborators can choose to ignore the rules and conditions that the composer has set for their collaborations or impose their own conditions on the work. As discussed with gamer motivations, the audience are going to behave in a way that will best suit their needs and requirements for the work and the work will be expected to adapt to these changing needs and motivations.[11] Rules shifting and adjusting through audience use has an impact on the amount of control the composer has over the work and the conditions for aesthetic completeness as this will be interpreted through the audience's lens in the absence of the composer. In contrast to Rohrbaugh's observations on the hierarchy of the artist,[12] for interactive music this effectively means that the only people who can decide if the work is aesthetically complete are the audience.

In the absence of the composer's judgement or decision on the completeness of the work, the audience will need to determine how they assess this aesthetic completeness based on their own sense of aesthetic values. These aesthetic values will be determined by their motivations for attending the work and what they wish to gain from engaging with the work, for example, they could be aiming to have a new musical experience, to feel that they have control over the music or to have an evening working collaboratively with their social group. Their sense of aesthetic completeness will also be impacted by their gamer motivation and how they approach a space. For example, somebody who approaches a space feeling like they need to achieve a sense of narrative completeness will feel like their experience is aesthetically incomplete if they have missed parts of the narrative.

An audience's view on whether a work is aesthetically complete is likely to be expressed through how they engage with the work and how long they continue to stay engaged with a piece of a work. The most likely indication from the audience as to whether they feel a work is aesthetically complete will be how long they spend engaging with and exploring the work before they finish interacting. If they finish their interaction after a long period of exploration and play feeling satisfied with the artistic outcomes and the meaningfulness of their interaction, then this can be considered aesthetically complete. If they finish interacting after a short period of time, do not feel that they were truly instigating the interaction in the work or do not explore and demonstrate playfulness in the work then the work could be considered aesthetically incomplete.

The composer's reliance on the audience to aesthetically complete the work requires an amount of trust. They need to be confident in the conditions that they have set for the work and how the audio interface will support the audience within their interaction. They also need to trust that the audience will approach the work with a willingness to interact and explore, treating the work on the terms in which it was created as a non-linear piece of music that requires their actions as collaborators to realise its full potential. This is where the establishment of liminal spaces become important for the composer to provide clear conditions for the work and interaction before the audience enter the space. This is a key tool for ensuring that the audience has all the required information to be able to aesthetically complete the work.

This is not unique to interactive sound and music. The completion of all sound and music requires some faith from the composer once the work is released into the world. For example, when a composer releases their music out in a commercial context, they are releasing a fixed recording with fewer variables than interactive music. Through this they can be relatively sure that the music is heard in the form that they intended; the mix will stay the same and the structure of the work will stay the same. They cannot, however, control the conditions that are in place when the audience member listens to the work.

Whether they listen intently or whether they use the mode of listening of 'distracted listening', where the music is on while the listener is engaged in other tasks like work or driving, is outside of the creator's control,[13] as is the audience's choice to listen through headphones, speakers or through a phone's built-in speakers. These are all aspects of quality control which cannot be determined or controlled in advance. The composer is only able to test the work sufficiently to ensure that the mix sounds as good on any system. Similarly, when a composer releases work in a classical context they are reliant on a mediator to interpret the work in performance, this may be a conductor or individual players. This may lead to variations in tempo, dynamics and musical interpretation. By releasing the music to the score the composer relinquishes some ability to fully control the music.

Interactive sound and music is no different in this respect. Once the composer releases the work out into its interactive context, they lose the ability to dictate the conditions of the work. While they can produce guidance, they cannot control how the audience engages with the work and the actions that the audience takes to interact within the work. They are releasing the work into the unknown where some audience approaches will be determined by social practices in different communities and by audience expectations. The composer cannot even be sure if the audience is choosing to engage with the musical aspects of the work as some game platforms have the option to switch off a soundtrack and players may take this option to replace the music with their own soundtrack. In interactive sound installations how audiences engage in group works will be impacted by group dynamics and crowd behaviours, and can be shifted by strong audience roles. If the way that a large group of audience members approach the exploration and aesthetic completion of the role changes, this can impact all other audience members within the work.

This again brings us to the question of control and collaboration in the audience and composer roles that was first discussed in chapter 1 when relating to concepts of authenticity in interactive sound and music and was further discussed in chapter 4 in relation to meaningful interaction in sound and music. A condition of a work being authentically interactive and providing a true collaboration from the audience means that the composer must accept that they will not have full control over the final output. For the composer this means releasing the work into the world and accepting that they will not hear the final result, only iterations of the music, and that the work will be at the whim of various audience behaviours. This is a key aspect to a person's role as a composer, at some point you have to give up your work to the world; what happens once it has been released is outside of your control.

Releasing the work to the audience to complete becomes like a case of endorsed fanfiction, where the audience are taking the work and remixing it and working with it on their own terms.

As Philip Pullman said when discussing fan interpretations of his work:

> When you're telling a story, you are the ultimate ruler. You have power of life and death over every comma. Every full stop every character every storyline, yes, and so it should be that's right and proper. So that's the totalitarian part of it the despotism. But as soon as it's written and published, and out there in the marketplace in the bookshops, your power vanishes, and it becomes democratic.[14]

As the composer has chosen to work in an interactive context, they have chosen conditions that mean that they will never see their work complete. They have chosen a medium that invites play and uncertainty and requires an understanding of their audience as true collaborators rather than participants. This is the strength and artistic challenge of interactive work.

The impact of versions in interactive music

For a piece of interactive sound and music where the composer has aimed for the audience to have achieved the role of collaborator, as outlined in chapter 1, the piece is never really complete until the work is installed with an audience or an instigator for the interaction. As previously established, each iteration of the work establishes a new approach to the collaboration and there may be many iterations that are never heard. In these cases, the work changes with each new audience or interaction with the work. Based on this, the composers are never able to hear the final iteration of the work and a final iteration of the work cannot theoretically exist as the work is forever shifting and changing. Even if a composer attended for the entirety of an installation, the work is loaded with possibilities and potentials as outlined in the phase space of chapter 3. They will not be able to hear every iteration as each version of the work will be different.

Definitions of completeness in arts are usually built from an understanding that once the work is released into the world it remains unchanged in a final version. Once the work is complete it is static and finalised. This contrasts with the examples set by practitioners and many artistic practices where a composer or artist chooses to revise their work at a later date. Artists may choose to revise a work based on a change in politics, changing acceptable standards for language or changing technologies that allow them to better realise their artistic vision. Rohrbaugh provides the example of George Lucas' reworking of the Star Wars films once new technology had become available to fully realise his intended effects.[15] Rohrbaugh's accounts of completeness allows for revisions that either represent a new version or a different 'complete' version of the work.[16] All these completed versions sit alongside each other rather than replacing the previous work. This also aligns with Davies' view of versions in music that include different instrumental configuration and minimal changes to the music.[17] However, these versions created through composer revisions stay solidified once they have been made, while in interactive sound and music the piece is constantly being redeveloped with different sounds and structures as the audience begin to take their role of the collaborator. There is never a solidified version that is released into the world.

Every time an audience or player engages with an interactive piece of music it will result in a different, new version of the piece. These versions are not revisions of the work and are not led by the composer, instead they are led by the audience collaborators. There is no one definitive version of the work, the versions are the completed work in multiple different forms. This aligns with Davies' approach to versions, where every revision of the work by the composer exists as a different but equally valid version of the work. Davies makes the condition that the change to the work comes after the completion of the piece and that the changes 'intentionally and moderately alter identity-relevant features of the original but without resulting in the production of a new but derivative work'.[18]

The difference between versions in interactive sound and music and in Davies' definition of versions is the voice of the composer. In an interactive context the composer lays the foundations for changes in the work, but they are not responsible for enacting them. Instead, they delegate this power to the audience as their collaborators.

This concept of multiple, valid versions of the work in the place of a definitive composer-led version links back to the open phase space that was established, a compositional process allowing for the composer to build into the compositional approach the variation and possibilities in their work that are required for an interactive piece. When an interactive piece of work is moved from the draft into the live performance space for interactive sound and music, where the audience is fulfilling their role as the collaborator, whether in a virtual game space or in a physical space, each iteration of the work as created by the audience through their interactions becomes a legitimate, complete version of the work with significant variations between the versions but not enough change that they could be considered a distinct new piece of work. The structure in each version of the interactive work will be formalised and the conditions for interaction will be set and unlikely to change. This means that with interactive sound and music there are an infinite number of complete versions of the piece that continue to evolve as the piece progresses, like a multiverse of completed pieces that expands every time an audience member deems the work to be ended.

Conclusion

When considering completeness in interactive sound and music, the role of both the composer and the audience needs to be taken into account as they are both equal collaborators in the work. While the composer is responsible for determining the genetic completeness of the piece where they have put in place all the requirements for a work to be technically complete and ready to release to the point where the audience can take over in their role as the collaborator, they cannot be held responsible for determining whether a work is aesthetically complete.

The audience, as collaborators in the work, is responsible for determining the aesthetic completeness of the work, ascertaining whether a work has succeeded in its aesthetic aims based on the point that they choose to finish interacting with the work and their perception of the aesthetic completeness.

Due to the nature of interactive composition processes the composer will never hear all the completed versions of their work. The amount of material and possibilities in an interactive piece will mean that anyone who engages with the work will only ever hear a version of the piece of which there will be infinite possibilities. The impact of this on the composer is based on how they view their role in the work. If a composer looks at their role as one of an auteur who has full creative control over the piece then this will have an impact on how they see the output. A composer taking this approach will see themselves as being responsible for all the music being created within the work and the quality of any outputs resulting from the piece. With meaningful interaction in the work, this becomes untenable for the composer. They cannot ensure the quality of all the musical outcomes without greatly restricting the amount of interaction within the work and, therefore, restricting the collaborative role of the audience in the work. A composer's need to control the work and the completeness of the output is directly in conflict with the audience's need and expectations for an interactive piece and their potential to create a meaningful collaboration with the composer.

For a composer that views themselves as collaborating at a distance with the audience, a sense of completeness in their musical work sits outside of their interactive and collaborative approach. The composer aims to create the conditions for the work but not a completed work in itself. This composer still needs to accept that there are versions of their work that will never be heard by themselves, or even by audience members.

In a sense, the concept of a finished work is not applicable to interactive sound and music. As the composition processes aim to keep the phase space open, the pieces being created will always just be one version of the possibilities, leaving the potential of infinite other combinations of works that exist and evolve every time an audience engages with the piece. It's this potential to keep evolving that represents the exact compositional challenge that makes the work appealing to composers and the individualised experience that is so appealing to the audience. With this in mind, the work can never truly be considered finished, only evolving.

Reading group questions

1   How can the composer communicate their understanding of when a piece of music is considered complete?
2   Should the composer be the only person to decide that a work is complete? What impact does this have on the hierarchy of composer and audience?

3   How does the way composition is taught influence a composer's sense of completeness?
4   If a composer cannot listen to all the possibilities within a piece, how can they make a decision on the completeness of the piece?
5   What does the composer lose by never hearing their completed work? What are the creative benefits to this approach?
6   Can an audience in the collaborator role truly judge the aesthetic completeness of the work? How can they be supported in this role?
7   How do fan fiction processes relate to audience collaboration in interactive sound and music? What can a composer learn from fan communities and how they engage with works of fiction?
8   At what point would a version of a work become considered a completely new piece?
9   With interactive work generating so many versions through each audience use, how do audiences engage with the music as a unified piece?
10   Could the concept of a finished work ever be applied to interactive sound and music?

Notes

1   Stephen Davies, *Musical Understandings and Other Essays on the Philosophy of Music* (Oxford: Oxford University Press, 2011), chap. 12.
2   Davies, pp. 177–78.
3   The implications of the role of the score will be discussed further in chapter 6.
4   Paisley Livingston, 'Counting Fragments, And Frenhofer's Paradox', *British Journal of Aesthetics*, 39.1 (1999), 14–23.
5   Guy Rohrbaugh, 'Psychologism and Completeness in the Arts', *The Journal of Aesthetics and Art Criticism*, 75.2 (2017), 131–41 (p. 131) <https://doi.org/10.1111/jaac.12370>.
6   Calculated as 10! (Factorial.)
7   Hatnote, *Hatnote Listen to Wikipedia* <http://listen.hatnote.com/#> [accessed 15 March 2024].
8   Karlheinz Stockhausen, *Aus Den Sieben Tagen*, 1968.
9   Some may disagree with me on this point, as the decision to follow the conditions dictated by Stockhausen would be the performer's choice.
10   Paisley Livingston and Carol Archer, 'Artistic Collaboration and the Completion of Works of Art', *British Journal of Aesthetics*, 50.4 (2010), 439–55.
11   Nick Yee, 'Gaming Motivations Group Into 3 High-Level Clusters', *Quantic Foundry*, 2015 <https://quanticfoundry.com/2015/12/21/map-of-gaming-motivations/> [accessed 12 March 2024].
12   Rohrbaugh, p. 133.
13   D. Huron, 'Listening Styles and Listening Strategies', in *The Society for Music Theory*, 2002.
14   'Philip Pullman', *The Adam Buxton Podcast*, 2019 <https://www.adam-buxton.co.uk/podcasts/25> [accessed 10 April 2024].
15   Although some may disagree with the effectiveness of these changes.
16   Rohrbaugh, p. 139.
17   Davies, chap. 12.
18   Davies, p. 179.

References

Davies, Stephen, *Musical Understandings and Other Essays on the Philosophy of Music* (Oxford: Oxford University Press, 2011).

Huron, D., 'Listening Styles and Listening Strategies', in *The Society for Music Theory*, 2002.

Livingston, Paisley, 'Counting Fragments, And Frenhofer's Paradox', *British Journal of Aesthetics*, 39. 1 (1999), 14–23.

Livingston, Paisley, and Carol Archer, 'Artistic Collaboration and the Completion of Works of Art', *British Journal of Aesthetics*, 50. 4 (2010), 439–455.

'Philip Pullman', The Adam Buxton Podcast, 2019 <https://www.adam-buxton.co.uk/podcasts/25> [accessed 10 April 2024].

Rohrbaugh, Guy, 'Psychologism and Completeness in the Arts', *The Journal of Aesthetics and Art Criticism*, 75. 2 (2017), 131–141, doi:10.1111/jaac.12370.

Yee, Nick, 'Gaming Motivations Group Into 3 High-Level Clusters', *Quantic Foundry*, 2015 <https://quanticfoundry.com/2015/12/21/map-of-gaming-motivations/> [accessed 12 March 2024].

Media and art examples

Hatnote, *Hatnote Listen to Wikipedia* <http://listen.hatnote.com/#> [accessed 15 March 2024].

Stockhausen, Karlheinz, Aus Den Sieben Tagen, 1968.

Chapter 6

Can we create an artefact of a piece of interactive audio?

Interactive sound and music is, by its very nature, difficult to pin down into one form. Composers and creators work hard to ensure that the work exists in many forms and that audience members or players get a different, individualised experience whenever they approach the work.

However, we are now at the point when elements such as the canon are being established and people are considering how work is saved and analysed in the future. Our current approaches are developed for musical work with fixed forms, which do not necessarily fit the needs of interactive sound and music, or any musical works that rely on more flexible forms and approaches. This creates a risk that we will not actually be creating an accurate artefact of the interactive work, but will instead be creating an interpretation of the work.

A thoughtful approach to creating artefacts from the work will support any future building of the canon so that it can include a range of works accurately preserved and displayed.

This chapter will address some of the concerns about archiving interactive sound and music starting with the concept of 'the work' when applied to interactive pieces. I do not intend to provide a solution for this challenge, but more address the precedents and consider a list of best practice considerations that consider interactive sound on its own terms, rather than using tools that have been developed for other formats. The chapter will examine the approaches for creating an artefact from interactive sound and music that consider the interactive nature of the work. This will look at the concept of 'the work' when applied in an interactive setting and consider potential solutions from a composition, analysis and archiving perspective.

Reasons to create an artefact and precedents

In April 2023 the 'Ground Theme', the main musical theme from *Mario Bros*, was the first piece of video game music to be entered into the Library of Congress, becoming part of their National Recording Registry.

At the point that the *Mario Bros* theme entered the archive I emailed the Library of Congress to request information on the considerations taken when

DOI: 10.4324/9781003344148-7

archiving the work. The work was provided directly as a .mp3 file from Nintendo, therefore it was considered that no further considerations were needed.

The Mario theme has become part of the cultural consciousness. YouTube covers of the theme include Marimbas,[1] orchestral covers,[2] live play along with games,[3] otomatones[4] and even card readers.[5] It is a culturally significant piece of work.

What is even more significant is the technology available when the piece was written and how it impacted the sound of the game.

As composer Koji Kondo stated in his interview for the Smithsonian:

> The amount of data that we could use for music and sound effects was extremely small, so I really had to be very innovative and make full use of the musical and programming ingenuity that we had at the time.[6]

Due to when the work was created, in the 1980s, and the limitations of video game technology at this time the *Mario Bros* theme is a looping linear piece of work. Therefore, storing the work in a recorded .mp3 format does not change any functions of the work but does change the context and sets a precedent for other game scores entering the archive.

Within the interview for the Smithsonian, Kondo emphasises the importance of interaction and audience behaviour to his work:

> I used all sorts of genres that matched what was happening on screen. We had jingles to encourage players to try again after getting a "game over," fanfares to congratulate them for reaching goals, and pieces that sped up when the time remaining grew short.[7]

This quote emphasises the impact that the technology available at the time had on the overall sound of the work, how the music was to encourage repeated plays (a hopeful 'game over' jingle) and to push the speed of the game. However, the storing and archiving of his work does not represent the gameplay aspect of his work and the audience roles.

The precedent set by the archiving of the *Mario Bros* theme is similar to the precedent set by live orchestral performances of games scores, which have become a staple of orchestral music[8] or soundtrack albums. By creating an artefact for archiving or for performance in a different form, a fixed version of the work is created and the context of the score is removed. The work becomes preserved in amber.

It seems as well that there is a perception that preserving games and interactive soundtracks in an orchestral format legitimises them as art as indicated through news coverage such as the 2011 article in *The Washington Post* which described the programming of a live orchestra performing games scores to 'show non-gamers how significant video games are',[9] or the London Video Game Orchestra which has a goal within their mission to 'increase the impact and value of video game music for our audiences'.[10] As Goehr states in *The Imaginary Museum of Musical Works*:

> The constant bid to define and redefine the concept of music derives from a need to convince the higher echelons of the establishment that certain musical practices are among those that are respectable and civilized.[11]

This move for traditional recognition of legitimacy can be seen across games discourse and discussions of games as art. In this case, it involves placing a format on the work that disguises the origin of the work itself, moving it to fit into a conventional concert hall using a linear format that does not represent the compositional considerations and complexities as discussed in chapter 3. In a way this is similar to an action figure that has been kept unopened in its original packaging. Yes, it has been preserved for future generations but hasn't it lost something of its purpose? For people to understand games and interactive music they need to understand the context that it sits within, rather than appreciating the melodies as an abstract idea in a different musicking context, where the audience are expected to sit still and listen to the work quietly rather than participating in creating the work.

Preserving interactive sound and music is a challenge. Essentially, we are creating an artefact of something that is supposed to be ephemeral, while, as previously discussed in chapter 3, the interactive composer is intending to keep the musical phase space open, allowing the audience to choose different options within the work. Linear formats (scores, recordings, videos) preserve one version and close the phase space, working against the composer's original intention.

Since there are so many challenges in creating an artefact from interactive sound and music a very valid question is why would we want to do this? If preserving the work loses the core of what the work is, why not accept that it will never be truly possible?

Some reasons for creating artefacts are practical for the composer:

- To recreate the work for the future. By their nature interactive performances are ephemeral and require instructions for set up to be staged in the future
- For a composer to advertise their work

These require detailed and realistic versions of the work demonstrating the construction and the sound in context. Through this the composer can ensure repeat performances of the work.

Some reasons are more esoteric and consider the cultural significance and future legacy of a piece.

These include:

- Providing a record of the work for analysis and study

- Preserving and archiving a work that is considered 'culturally significant' (as is the case with the Mario theme)
- For future audiences to experience what happened, this will potentially be at a point where the technology used to create the original work or the locations that the work was experienced in are no longer available

We are in an interesting position when considering preserving and archiving work that is considered 'culturally significant'. With interactive sound and music being a field that only began developing in the late 20th century, we don't actually know yet what is culturally significant. The canon has not been established and we do not yet have the requisite distance to look at the importance of different pieces of work, but we are attempting to start this process, particularly with games. Because of this there is a risk that we won't have preserved files in a format that allow for the significance of the work to be appreciated in the future. This is a situation that we have found ourselves in when preserving artefacts reliant on technology, for example in the sound archive at the British Library where some of the recordings are deteriorating quicker than they can be preserved.[12]

While technology is a key consideration when creating an artefact, the bigger precedent being set is the musical meaning of the work and ensuring that the interactivity and context of the work being preserved remains intact where classical or popular music approaches to music are being used as the default for storing all forms of music.

Goehr raises this in relation to scoring and performance practices in the Romantic period: 'one way to bring music of the past into the present, and then into the sphere of timelessness, was to strip it of its original, local, and extra-musical meanings'.[13]

Through the practice of storing interactive work as single recorded files or as scores, we risk repeating this approach and removing the context from interactive sound and music.

Through the Mario Library of Congress example, and the orchestral live performance approaches, a precedent has been set to create artefacts using traditional classical methods which move the music to a singular version laid out in a score or a recording. This precedent suggests the following:

- There is one 'true' version of a piece of interactive music. This is the one that is the most melodically 'complete'
- The music exists away from the concept of its use, so a game theme can exist on its own away from the gameplay context

However, as previously discussed in how the compositional process changes in interactive music, this doesn't fully represent the process by which the work is constructed. By adapting the work to represent a 'true' approach without loops or variations, two things are suggested:

1   The composer was always aiming for a 'complete' version of the work or the composer composed the work in its entirety and then retro-fitted the interaction through editing (this may be the case for some composers)
2   The interaction gets in the way of the concept of 'the work'

This does suggest hierarchies of legitimacy in music, that all composers are aiming for an ideal in their work based on Romantic approaches. However, there is a particular skill in working in an interactive format and being able to conceive of multiple versions of a work and this should be represented in any artefact being created. While we have previously discussed interactive sound and music within the context of a phase space, allowing the composer to keep open paths and opportunities, the act of archiving can solidify a path through the work which closes off the other spaces and moves the work outside of the composer's intention for exploration and change.

The musical work in interactive music

At the centre of the precedent of how interactive sound and music has been preserved is a decision on what actually is the core of the music that we are preserving, linking to the concept of the 'work' and bringing into question what can be considered the 'work' in an interactive setting. Discussions about the work in an interactive setting build on long standing existing debates in music about the concept of the 'work', a flawed but appealing concept as it adds a weight and legitimacy to the importance of a piece of music, especially in a new and developing field such as interactive sound and music where we are essentially in the process of defining the canon, or an equivalent.

In the introduction to *The Imaginary Museum of Musical Works*, Lydia Goehr outlines the challenge of the 'work' as a concept:

> Musical works enjoy a very obscure mode of existence; they are "ontological mutants". Works cannot, in any straightforward sense, be physical, mental, or ideal objects. They do not exist as concrete physical objects; they do not exist as private ideas existing in the mind of a composer, a performer or a listener; neither do they exist in the eternally existing world of ideal, uncreated forms. They are not identical, furthermore, to any one of their performances. Performances take place in real time; their parts succeed one another. The temporal dimension of works is different; their parts exist simultaneously.[14]

I would argue that the 'work' becomes even more challenging as a concept when we consider it in an interactive framework. If traditional forms of music have ephemeral contexts where they exist in the mind of the composer, through performance, through the audience listening and through recordings and notation, then interactive sound and music exists in those formats but in multiple varying versions that change based on audience behaviours as seen in the phase space of composition. To complicate this even further, there are versions of the music that exist only within the actions of the audience and change with every audience or player that interacts with the music. Within her writing Goehr discusses whether the term 'work' should be applied to styles of music outside of classical music as this imposes certain approaches on the music, such as scores and recording practices.[15] While not necessarily created with other compositional styles in mind, the term 'work' does get applied to other contexts and musical styles. In interactive sound and music we use this term in certain contexts, particularly when describing interactive sound and music that exists separate to other types of media (e.g. games) such as sound installations or public art. Therefore, we need to establish what is meant by the 'work' in such a changeable field. While imperfect as a concept, and very much based in a classical framing, it is important to consider the work when we are thinking about what exactly it is that we are preserving when we archive interactive sound and music. Considering what is the work will help to define priorities and the essence of what is needed within the artefact.

While Goehr has mentioned the 'temporal dimension' of music, interactive sound and music shifts that temporality. As outlined in chapter 3, while there are linear aspects to the construction of narrative-driven interactive audio, the work is no longer fully linear, and sections may be heard in varying orders. Through the use of moment form, we also remove the hierarchy of some linear structures which removes the grammar by which we usually understand and interpret music. Additionally, the role of the audience adds more complexity to the challenge, the variations are determined not by one performer or improviser acting as a mediator for the composer, but by audience interactions. So while traditional music forms exist in multiple, simultaneous formats experienced in real time and stored through the score, interactive sound and music also exists simultaneously in a potential format where the music exists but other choices have been triggered within the piece, as a phase space.

For many people the score is the truest representation of music and represents the 'work'. People will bring scores to classical performances and read along, a practice that is very disconcerting if you are performing as it leaves no space for small errors or differences in interpretation. A visit to the British Library Treasures exhibition will reinforce the viewpoint of the score as the music. Displayed are a number of varied scores from the renaissance period up to graphic scores. In some cases where scores are not the standard practice, e.g. in popular music, lyrics are included with accompanying recordings in listening posts. In contemporary popular music, the music may be considered the recording as the studio becomes a compositional tool. The live performance exists as a different interpretation.

As Busoni states:

> Every notation is, in itself, the transcription of an abstract idea. The instant the pen seizes it, the idea loses its original form. The very intention to write down the idea, compels a choice of measure and key. The form, and the musical agency, which the composer must decide upon, still more closely define the way and the limits.[16]

As Goehr raises, classical styles have developed a performance approach based on a perceived attempt at perfection, how closely can the performer replicate the score? This approach is based upon an assumption of composer's intention, if the score has recorded the composer's ideal then this is what the performance should be aiming to replicate completely. However, this doesn't always take into account the performance styles of the time or assumed techniques that were so prevalent that the composer did not feel a need to express them in the score.

The very nature of interactive work means that we are not aiming for a consistent performance, as some forms of classical music do. There is no score guiding the performer to create one version of the work that measures up to an 'ideal' performance. In fact, creating a score of the work adds to the problem as it notates one instance of the piece, preserving that as the piece itself. Based on the amount of variations possible and the transient nature of interactive work, a fair comparison might be with jazz and improvisation. In jazz there exists multiple variations that exist in the moment. Due to its ephemeral, improvisatory nature we accept that jazz cannot be fully preserved; this becomes a feature rather than a bug and any recordings preserve one version of the music but do not provide a definitive 'work'.

In jazz a different standard is set based on concepts of authenticity and the ephemeral where the imperfections in improvisation are not something to be fixed but represent the work being live and in the moment. This is laid out by Hamilton in the *Aesthetics of Imperfection* where Hamilton defines imperfection as involving:

> ...an open, spontaneous response to contingencies of performance or production, reacting positively to unpromising as well as promising circumstances. It can include idiosyncratic instruments, apparent failings in one's performance or that of colleagues, and age and infirmity.[17]

In imperfection we potentially find a better framing for interactive sound and music where the composer needs to relinquish control over the work in order for it to be fully realised. The goal is not for every audience member to have the same experience every time, but for individualised experiences. Leaving space for imperfection allows for the work to reach its true form.

Hamilton states, when debating the supposed conflict between process and product:

> The *artwork* does not have to be closed; process and product are two sides of one artistic coin. "Processual" means "highlights or foregrounds the process of production"; a non-processual product attempts to conceal that process, but a processual one does not.[18]

In interactive sound and music we are attempting to keep the process open as part of the final product. While some technology may be concealed the composer is not hiding that a process is taking place; we expect the audience to become part of the process. Therefore, when creating an artefact of the work, we need an approach that facilitates the process as well as displaying the final outcome.

When considering jazz, we have reached a convention that uses the following:

- Lead sheets that notate the head with chord patterns for improvisation
- Transcriptions of improvisations
- Recordings of versions

Though it is generally accepted that the ephemeral nature of improvisations is part of the music itself so it can never fully be preserved as a 'work', the nature of recording multiple versions of each performance and improvisation does lead to a collector's approach where fans will look for recordings of different improvisations, building a picture of the artist through different performances and determining which improvisatory tools stay consistent through performances.

This collector's approach can exist in interactive sound and music, as outlined through Yee's gaming motivations[19] where a player motivated by achievement might repeat gameplay so that they've unlocked all possible variations in a work, or in immersive theatre such as Punchdrunk where an audience member might return for multiple viewing to ensure that they have gained all possible experiences in a theatre production.[20]

While the ephemeral nature of jazz improvisation is consistent with interactive sound and music and the understanding that every performance may be different with shared consistencies across visits, the other conditions held within jazz may not be directly applicable to interactive sound and music. Jazz still works within linear structures, and this time-based approach does not translate over to the more installation or immersive styles of interactive sound and music that work within moment form structures. Additionally, elements of reproduction and interpretation in a live context can be considered different when comparing jazz with interactive sound and music. In jazz there are multiple versions of the work that are performed and interpreted by different artists and instrumentation. In many interactive structures, such as sound installations or video games, the instrumentation and style of the work are fixed by the composer within the technology before being mediated by the audience, in a role similar to the performer.

When considering the work, a more accurate comparison for interactive sound and music may be with avant-garde practices such as those used by Cage in his aleatoric work such as *Music of Changes*,[21] which is built on random change using the I Ching, or *Variations III*,[22] where performers drop circles of events onto acetate to determine the musical actions.[23]

As Goehr raises, these compositions were developed in part to reject the traditional, Romantic concepts of music but do not fully reject the concept of the work in a way that could be seen as directly applicable to the forms of interactive sound and music:

> Cage and Stockhausen have sought ways to undermine their control over both performance production and their compositional procedure. If the "work" itself can be formed as a result of as little controlled interference as possible, the resulting music is as close as it can be to being uncomposed or unintentional.[24]

This is particularly applicable for composers coming from experimental approaches to interactive sound and music and those working in sound installations. These types of work link themselves in form to the avant-garde and, as shown in chapter 4, build on experimental and classical traditions. Additionally, these indeterminate works reduce the control of the composer and bring performers into the role of collaborators, similar to the role of the audience within an interactive work. As Goehr outlines, composers within the avant-garde use the term 'work' interchangeably based on different contexts to their benefit.[25] This can be a political choice for work to sit within the academy or to act as a challenge to the academy but can also allow for the music to sit within different communities. Using the term 'work' suggests a Romantic framing, and eschewing this may be a choice to deliberately make a piece fit more within a public work or theatre context. Very practically speaking the term can also used for convenience as 'work' has become standardised within the field.

This suggests that, similar to composers working with indeterminate structures, the composer of interactive sound and music can self-define with the term of the 'work', applying it to a range of structures as relevant to the music that they are creating.

However, a large difference when considering more avant-garde approaches is the central role of the score. Many of the pieces working with more improvisational or chance structures in experimental composition provide a score that records the process as the intention for the composer and audience. Even Cage's *4' 33"* has a score that indicates how to perform the work, including how to indicate the differences between movements.[26] Due to the background of the avant-garde subverting traditional practices this creation of a score, even within more creative formats, does make sense for preserving the work but does make the practice different to the needs of interactive sound and music, which very rarely includes a formal score beyond sketches. In interactive sound and music, outside of avant-garde contexts such as in games or in public art, scores very rarely exist and do not provide a useful context for the work. In these cases, the interface and instructions embedded within the work act as a guide for the audience or players to understand the expectations for interaction.

Precedent for analysis from games audio and ludomusicology

Creating an artefact of the work is necessary to create a record of the sound and music for promotion, analysis and for your own records of your work. The standard process for this is to create a video walkthrough of a project. However, this just demonstrates one version of the work and one set of decisions showing a fraction of the work created and the phase space, meaning that it cannot be considered at all representative of the piece.

When considering video game music, a number of attempts have been made to preserve the work for analysis. Tim Summers has provided a comprehensive guide to analysis within the book *Understanding Video Game Music*.[27] Within Summers' approach the aim of the analysis is important, e.g. whether the analysis is investigating the audio implementation, melodic and musical features or impact on gameplay. The approach used is adjusted based on these contexts.

Summers outlines an approach based on analytical game play that demonstrates the interactivity of the game and recording multiple game sessions so that the player/analyst can observe changes in music based on different forms of gameplay which can be considered similar to the approach outlined to testing a composition as outlined in chapter 3. In general, there are multiple approaches to playtesting, but the player is looking to emulate different forms of play aligned to player motivations as outlined in chapter 2,[28] but they are also looking essentially to break the game to understand how it works. For example, a standard trigger for audio in a game is a collider, where the player entering or exiting the object acts as a trigger for audio to start or stop. By repeatedly walking across a threshold in a game and observing the audio the analyst can determine what types of triggers are used in a game and where they are placed. A musicologist may also look for 'seams' in the construction for the audio such as repeating sounds or gaps at the end of loops to help them understand how the music was written and how it adapts to the space. Other speed running processes can be used to test how a game was constructed. For example, designers put in shortcuts through a game to allow for them to test changes at speed. Finding these shortcuts, if left in the game, can support understanding of how a game is constructed.

Summers also advocates for an understanding of software that is used to build games such as the game engines and audio middleware and basics of coding. Here the platform used for the game becomes important. If analysing based on a PC game, code stored in text files is more readily accessible and can be used to understand how the game works. Further to this, if we use the collider example above, understanding of audio implementation from a practitioner's perspective can be useful tool for analysis. For example, if you understand how a collider is added into a game and how game states are used when implementing audio, you can look for these within a game when playtesting and analysing the work. From here, having built a picture of the possibility of game music, Summers repurposes many approaches that are standard to musical analysis, including motivic, melody and harmonic analysis as well as semiotics and topical analysis, while adapting these forms to acknowledge the non-linear structure of games.

What this approach to analysis does is preserve the multiple possibilities of the work, while constructing the analysis within an appropriate technical context. It is not imposing linear traditions on the music but is looking at a number of different versions.

The context of a work is also important within this style of game analysis and Summers does also advocate that analysis includes sources outside of gameplay including press material, interviews, game maps etc. to provide additional understanding and context of the construction of the game, the audience and how creators expected the user to interact with the work.

This comprehensive approach provides a strong oversight for research of gameplay, sound and composer intentions which does not rely on preserving vintage consoles. It does rely on a very technical player/analyst who is able to understand existing structures of games and audio implementation and does benefit those who also have a practice-based output in interactive audio who are able to interpret the technicalities of the game files.

This approach to analysis through developing multiple artefacts can be applied to wider examples of generative or interactive sound and music. For example, if you take a site such as *Listen to Wikipedia*,[29] an online installation that creates a generative algorithmic composition based on live Wikipedia edits, you can apply a similar set of actions. The piece works by different actions generating different sounds, with different instruments or timbres being applied to data being added or removed from Wikipedia and a chord playing for each new user. The pitch of the note is dependent on the size of the edits. Listening to and recording different versions will allow for a compositional analysis which establish the key and parameters in the work, while the explanatory text on the website used to label the visuals explains the meaning of choice of instruments and notes as well as what has been edited on Wikipedia (providing confirmation that the data is being updated in real time). Further to this, right clicking on the window that the installation sits within and selecting 'view page source' will show the code underneath the visualisation, including source sounds, and the corresponding Github pages allow for you to see further into the source material and the algorithm, which can confirm what was established through listening while providing an insight into the wider range of compositional options available in the work.[30] This transparency of material is commonplace in more experimental interactive sound and music practices that build on experimental approaches and open-source ideals of organisations like CCRMA, through this many composers make files and patches readily available online[31] which can support analysis.

Having wide access to source material and files and building an understanding of the composer's intentions for analysis relies on artists and composers subscribing to the values established in experimental interactive sound and music as outlined in the introduction where open-source approaches are preferred. However, interactive installations or public art remain even more ephemeral in nature. This is due to the role of site in many interactive pieces of work, the time limited nature of the display and the intended audience. For example, works displayed at a public art festival, such as Artichoke's *Lumiere Durham*,[32] require people to visit the site and experience the work in person. The site and community experience are essential to how the work is experienced and it is expected that the audience will be the general public rather than other composers and developers. Additionally, the works are displayed for only four nights, and the events are often crowded due to the popularity of the festival.[33] This means that repeated viewing and experimentation, as can be used for games analysis, will not necessarily be possible, especially with the large amount of attendees at the event. With time restraints on the number of performances it certainly would not be possible to undertake a walkthrough in all the potential audience modes available.

In public sound installations the technology and source files are often concealed, preserving the audience experience but also protecting the technology from the elements or from audience damage. While it isn't necessarily the intention of the composer to create a technology black box, it is key to audience immersion. Hiding the technology can lead to a sense of wonder when interacting with public sound work, for example, if you look at Daily tous les jours' *21 swings*,[34] part of the magic lies in the surprise of the soundscape. This is established as a guiding principle for Daily tous les jours: 'we believe that magic can only happen through the illusion that technology isn't there. But it is, invisible, hidden within the installation to better serve the interaction'.[35]

This can mean that within the nature of public art or sound installations looking at the code and technology may not be instantly possible, as the audience types and expectations override the need to make the tech visible.

In this case an alternative solution is offered that addresses Summers' approach for analysis through different audience type; crowdsourcing the walkthroughs for multiple audience members may provide an option to see how the work reacts based on different audience approaches to a space and interaction types. This has the added benefit of being able to cover a large amount of area in a single event. An example of this in practice can be seen in Atkinson and Kennedy's research into Secret Cinema, where student researchers were enlisted to attend a Secret Cinema event and report on their experiences.[36] Through this they were able to build an understanding of different participants' experience within the space.[37] This does require a consistent approach to tester documentation of their experience, including some awareness of how they approached a space, their expectations and the actions taken within a space.

The analytical approaches outlined are a practical way to understand approaches being used in interactive sound and music, but do not fully address the intentions and the feel of the work as it was originally displayed and experienced by the audience. The necessary need to reduce to constituent parts for analysis can put people at a distance from the composer's intention or approach from a practice-based perspective. With that in mind, a composer may want to consider how much of their methodology they share within supporting documents and with the availability of the files to ensure that their creative approach is considered within analysis.

The composer's intention

The composer's intention is a key consideration when creating an artefact of a work as this provides contextual information about the role of the work, intended audience and other additional cultural considerations. Contemporary interpretations of the work in classical music are often underpinned by a Romantic *werktreue* concept which better links performance to historic practices, for example ensuring that historically accurate tuning or instruments are used. Within *werktreue* framings the goal to fully realise a work is to get as close as possible to a composer's original intentions for a work, aiming towards a perfect realisation of the music.

In interactive sound and music the composer's intention involves creating a piece of work with multiple versions existing within a phase space and collaborating with the audience, relinquishing control of the work meaning that there is no one perfect realisation of the music, but many options. With this in mind, *werktreue* provides an interesting paradox when creating an artefact from interactive sound and music. In order to fully represent the composer's intentions and aims for the work the artefact needs to show the amount to which the composer did not exercise control of the work. Any archiving decisions or decisions around artefacts need to represent not only the composer's intention but the audience's input. Essentially, any decisions made around creating an artefact need to show either an absence of control or a potential for collaboration, two things that do not exist in a fixed and easily demonstrable form. This may be why people default to a single solidified version of the work because how do you demonstrate an absence within an artefact?

However, it is a challenge where a workable solution is needed. Interactive work cannot exist without the audience. If an artefact that has been created removes the audience's role from the work, suddenly it has a different purpose to the composer's intention. Again, this provides links to practices in jazz where the performer/improviser holds an important role in the music as they take on a compositional role through improvisation essentially making them a collaborator in the performance. However, the role of the audience is much less predictable than a performer/improviser and to truly preserve the interactive work that unpredictability needs to be represented in any archival decisions. If, when creating an artefact from an interactive piece of sound and music, we are intending to replicate the composer's intentions then we need to retain the interactivity and spontaneity present in the original work including the unpredictable audience behaviours. Therefore, any artefact created should not solidify the form of the work and should not exist without audience input. Theoretically, the ideal of any artefact created from an interactive piece of sound and music should also be interactive.

Further to this, interactive sound and music is usually built with further contexts in mind, including applied audio in games where the audio acts as part of the interface, community, public art contexts or site-specific contexts. Moving to an artefact needs to take into account these contexts as an extension of the composer's intention for the work as these will have been considered within the development. Any frameworks that are developed should represent this contextual information.

To be truly representative of the composer's intentions, any artefact being created for interactive sound and music needs to fulfil the following conditions:

1   The artefact should represent the multiple possibilities in the work, not a solidified version of the music. This should keep the phase space open and fully represent the amount of material that has been generated for the interactive work
2   The artefact should represent the context in which the work was created. This includes representing how the audio was intended to be used, aspects of musicking and the location for the work
3   The artefact should represent the audience or player roles within the work as essential collaborators to the work ideally through enabling interaction

These conditions create a number of practical challenges when creating an artefact from interactive sound and music.

Firstly, if we are creating an artefact for the future that fully represents the interactive nature and the context of the work then technology will be at the centre of any solution. Any file types will be specific to the technology on which the work was made, which may not exist in the future in order to recreate the work. Due to the speed of technological changes, and inbuilt obsolescence in some technology designs and the bespoke nature of some technology used for installations, the easiest solution is to store the original technology used for the interactive work.[38] However, this requires a lot of storage space, upkeep and technical knowledge. As an alternative, emulators can act as a cost-effective approach to run the original files and recreate the experience on contemporary hardware, but this will not always be possible for some site-specific sound installations or pieces of work with bespoke hardware, particularly if external sensors are needed as this hardware will require additional storage and upkeep. Additionally, an emulator approach will not necessarily be able to represent the community experience of an interactive sound installation if it cannot be displayed or recreated in a community context. Further to this, to preserve all the opportunities existing in an interactive piece of work requires a large amount of digital storage space, especially if the work is built from pre-recorded sound files. This does affect individuals storing records of their work who are reliant on external storage (which again is technology that can become obsolete). This storage will also have an impact on future archiving decisions, especially when we're still at the point where we do not know what pieces will be considered to be of cultural significance in the future. Digital storage space can be expensive to run and decisions on cultural significance may need to be made earlier than we can fully ascertain the importance of a work.

VR and XR as an answer

To represent the composer's intention through the original audience or player experience and context of the work it is necessary to consider not just the experience of the individual but also shared experiences of community. This is particularly important to be able to represent musicking within the context of the interactive work. Representing the community experience means that either the artefact will need to be usable by multiple people simultaneously or be able to accurately represent a shared experience to a lone user. Additionally, representing the site-specific and tactile nature of work will require either the storage and recreation of sets and materials or representation of the physicality of a site including touch-based experiences.

VR can offer an interesting solution for preserving an interactive sound installation, especially one with a site-specific context. Through tools such as LiDAR scans and game engines, the composer can recreate the space in a digital realm, offering the audience an accurate representation of how the work was staged and functioned. This became an expanding practice during the pandemic to represent museums or tourist locations that were closed during lockdowns, and could be expanded to other experiences. In particular, Google Street View has a range of virtual experiences on its platform including representations of closed galleries that can be experienced while no longer open in museums.[39] The representation of space in VR can even extend to accurately replicating the feeling of the sound in the environment through placement of speaker objects and the use of convolution reverb to model the acoustic of the space. As the technology develops this is potentially the closest way to create an accurate facsimile of an interactive sound installation.

However, currently there are technological limitations to VR and XR experiences meaning that some aspects of the work cannot be represented in a digital context. Firstly, the tactility of the work cannot currently be represented within a VR context. There is ongoing research into haptic technology to represent touch within a space but this is not currently able to represent the materials and weight of spaces and touch experiences so any touch interaction will be different to the real-world equivalent.

This is particularly important when considering the types of material being used in interaction as anything soft or fabric-based cannot be replicated with plastic or glass surfaces that are common in VR equipment. Furthermore, movement around a space is currently limited in VR environments, as users are anchored in one place through wires, and there is some danger in moving around a place with VR headsets in their current iteration where the user cannot see their environment (although this might improve with the range of XR tech being released such as Apple's Vision Pro). Additionally, there are some users that cannot fully use VR due to latency issues causing travel sickness, or who are unable to focus the images without having double vision. While there are investigations into walking platforms for VR to better represent movement, this is still in development and not readily accessible for all users so many VR experiences rely on a teleporting motion to represent movement which does not fully represent the experience of physically exploring an installation. Similarly, while convolution reverb can be used to create a sense of space this is still mitigated through headphones which can close off the sound making the audio experience feel artificial.[40]

Providing a VR rendering of a work also does not fully consider the social function of the work. VR is still by and large a solo activity, even if developments in the metaverse are building towards more socialization in the VR realm. The way that VR headsets are currently designed means that you cannot physically interact with other people within the room and are physically blocked off from seeing and interacting with people within a space. A solution would be to move the community interactions entirely to an online provision with users communicating through the virtual space, but any virtual interaction will not hold the same physicality as meeting in person. AR and XR may provide a solution through allowing an installation or sound work to be projected into a space, with users being able to view and experience the same work in real-time together. However, this will reduce the experience of the space that can be emulated in VR and still cannot represent the tactility and weight of the space.

These solutions using VR and XR are currently in early forms and may develop further as the technology develops providing the most accurate way to represent an interactive sound installation within a space. While space can be accurately represented, until the community experiences of interaction and the tactility of being in a physical space can be shown these will still be limited in how accurately they represent the experience.

Conclusion

While precedents are still being set about archiving and creating artefacts from interactive sound and music, we need to work with assumptions about what might be needed and useful in the future. The approaches that have been taken to date, as represented by the inclusion of the 'Ground Theme' in the Library of Congress, set a precedent that may have repercussions for future understanding of interactive works. If we move to using audio recordings of interactive work, with no further contextual files and information, as our main form of artefacts for interactive sound and music, when future generations come to study or recreate work they will not have access to a large amount of musical material within the work or additional information that supports understanding of the context of the music.

The concept of the work within interactive sound and music is complex but needs to be established to determine how interactive sound and music is preserved and what is considered part of the work. The use of this term by composers changes often based on context, with composers working in an experimental sound and music context using the term of the 'work', linking practices to more classical context, while applied sound and music, such as audio for games, being referred to within the context of the wider media.

A work within an interactive context is developed with a variety of contextual information about audience roles within the work, including unpredictable audience behaviour that needs to be represented within any artefact that has been created. Additionally, further context in the work can depend on the tactile information and the setting of the work, while aging technologies can make creating an artefact out of the work, that can be preserved for the future, particularly challenging.

Based on these challenges, the answer for creating an artefact of the work must be dependent on the context and the aim of the artefact:

To recreate works in the future: The artefact must provide not only the files, but a blueprint to recreate the work within an appropriate context (e.g. an appropriate location). This should include high quality files and documentation of how to assemble the work, including access to the technology from the time that can be used for future performances. There is a potential limitation to this practice where if the technology breaks and cannot be restored, the work can no longer be replicated. In this case, it is practical to also include alternative options including VR emulations of the work.

To provide a record for analysis: The artefact should provide a clear picture of the material in the work and the construction of the piece in order to allow for source material to be fully investigated. This can be provided through a combination of multiple videos showing different player behaviours, emulations of the work, source files and diagrams. Where a piece is large, such as in a large-scale installation or immersive production, this collecting of experiences can be outsourced to multiple audience members or researchers.

For future audiences to experience what happened: Similar to providing an artefact for analysis, the artefact should provide all the information needed to provide an accurate picture of the interactive work and the context in which it was staged. This should include videos, emulations including VR recreations, audience accounts of the experience, source files and diagrams.

To advertise the composer's work: Similar to allowing future audiences to experience the work, this should provide an accurate picture of the work and the audience experience, but it can be less detailed and provide fewer iterations of the work. This should include videos and emulations and could include VR recreations to show the experience.

However, a much larger question is should we be creating artefacts of interactive sound and music at all, or should we accept the ephemeral nature of the work as a feature which cannot be saved or archived for the future?

The appeal of interactive work is that it is ephemeral, the music can change and the audiences can be unpredictable. These works are often aiming for individual audience experiences that can't be recreated, and these personal experiences are part of the appeal for audiences to attend. Many immersive and interactive experiences now ban phones and cameras from the experience, to encourage people to be in the moment and as a feature for audiences to accept that they will have no record of the work after they leave.

Due to the interactive nature of the work, it is difficult to truly represent the phase space existing in interactive composition alongside all the contextual information, community interaction and the tactility of the work, meaning that something will always be missing from the artefact. As John Cage said, 'let no one imagine that in owning a recording he has the music'.[41]

By preserving any form of an interactive work, even with considerations made for context and audience experience, we are directly working against the nature of the work. By attempting to trap the work within an artefact, we are only storing a representation and cannot expect to capture the full contexts of an interactive piece. If we are building on avant-garde contexts, we must accept that an artefact can never represent the work and that the experience exists as only a moment in time, and that is part of the joy of interactive sound and music.

Reading group questions

1   What approaches can a composer take to make their technological approaches transparent while still preserving the immersion of the work?

2   How does how the concept of the music 'work' is used in an avant-garde concept affect our use of the concept within interactive sound and music?

3   How important is the composer's intention when creating an artefact of the work?

4   How important is knowing about audience interactions to understanding an interactive piece?

5   How much do tactile elements contribute to the experience of interactive sound and music?

6   How could you represent collaborative experiences in interactive sound installations within an artefact?

7   Within an interactive piece of music, what would represent perfection or imperfection? How does this impact choices when representing the work?

8   How does the social experience in the room impact the use of a VR experience?

9   How accurately do the virtual museum experiences represent the experience of attending a museum? How does this impact understanding of building an artefact from interactive sound and music?

10   What would be lost if we chose not to create artefacts of interactive sound and music?

Notes

1 *Super Mario Bros. on Marimba (with 4 Mallets) by Aaron Grooves*, dir. by Aaron-Grooves, 2015 <https://www.youtube.com/watch?v=3v7QC6Bl_7E> [accessed 15 March 2024].
2 *Super Mario Bros. Symphony Performance | Game Awards 2020*, dir. by GameSpot, 2020 <https://www.youtube.com/watch?v=lIS9alKXvz8> [accessed 15 March 2024].
3 *Let's Play LIVE! #1 - Super Mario Bros. w/FULL ORCHESTRA! Ft. George Salazar*, dir. by The 8-Bit Big Band, 2018 <https://www.youtube.com/watch?v=2cRPzzZV8u0> [accessed 15 March 2024].
4 *Super Mario Bros. Theme - Otamatone Cover*, dir. by TheRealSullyG, 2020 <https://www.youtube.com/watch?v=gjDJjI82xtc> [accessed 15 March 2024].
5 *Super Mario Bros. Theme on 2 Credit Card Machines*, dir. by Device Orchestra, 2019 <https://www.youtube.com/watch?v=2NCviYMdV1I> [accessed 15 March 2024].
6 Teresa Nowakowski, 'Super Mario Bros., Madonna and More Join the National Recording Registry', *Smithsonian Magazine*, 2023 <https://www.smithsonianmag.com/smart-news/national-recording-registry-mario-bros-madonna-180981988/> [accessed 15 March 2024].
7 Nowakowski.
8 Examples include the VIDEO GAMES IN CONCERT series held at the Royal Albert Hall where the 2023 production included pieces from games such as *World of Warcraft, Hades* and *The Witcher II*. Royal Albert Hall, 'Video Games in Concert' <https://www.royalalberthall.com/tickets/events/2023/video-games-in-concert/> [accessed 15 March 2024].
9 Moira E. McLaughlin, 'Video Game Music as Art?', *Washington Post*, 21 July 2011 <https://www.washingtonpost.com/lifestyle/style/video-game-music-as-art/2011/02/07/ABYW1SI_story.html> [accessed 15 March 2024].
10 London Video Game Orchestra, 'London Video Game Orchestra' <https://www.lvgo.co.uk/our-mission> [accessed 15 March 2024]; the question of 'legitimising' the form does sit outside the remit of this particular discussion but will be touched upon later in the chapter when looking at the concept of a music 'work'.
11 Lydia Goehr, *The Imaginary Museum of Musical Works: An Essay in the Philosophy of Music* (Oxford: Clarendon Press, 1992), p. 70.
12 Andy Linehan, 'British Library: Saving Our Sounds | British Council Music' <https://music.britishcouncil.org/news-and-features/2023-01-30/british-library-saving-our-sounds> [accessed 25 March 2024].
13 Goehr, p. 246.
14 Goehr, p. 3.
15 Goehr, chap. 9.
16 Ferruccio Busoni, *Sketch of a New Esthetic of Music* (New York: Dover Publications Inc., 1962), p. 85.
17 Andy Hamilton, *The Aesthetics of Imperfection in Music and the Arts: Spontaneity, Flaws and the Unfinished*, ed. by Andy Hamilton and Lara Pearson (London, England: Bloomsbury Academic, 2020), p. 31.
18 Hamilton and Pearson, p. 40.
19 Nick Yee, 'Gaming Motivations Group Into 3 High-Level Clusters', *Quantic Foundry*, 2015 <https://quanticfoundry.com/2015/12/21/map-of-gaming-motivations/> [accessed 12 March 2024].
20 Meghan O'Hara, 'Experience Economies: Immersion, Disposability, and Punchdrunk Theatre', *Contemporary Theatre Review*, 27.4 (2017), 481–96 <https://doi.org/10.1080/10486801.2017.1365289>.
21 John Cage, *Music of Changes*, 1952.
22 John Cage, *Variations III*, 1962.

23 N.B. 'Musical actions' has been used here in place of sound or music as the piece does not include any instructions for producing the sound and does not reference 'sound' within the instructions.
24 Goehr, pp. 263–64.
25 Goehr, chap. 9.
26 John Cage, 4' 33", (Original Version): Other Variations (London: Edition Peters, 1952).
27 Tim Summers, *Understanding Video Game Music* (Cambridge University Press, 2016).
28 Yee.
29 Hatnote, *Hatnote Listen to Wikipedia* <http://listen.hatnote.com/#> [accessed 15 March 2024].
30 'Hatnote/Listen-to-Wikipedia' (hatnote, 2024) <https://github.com/hatnote/listen-to-wikipedia> [accessed 18 March 2024].
31 The IRCAM forum and cycling 74 forum are both excellent places to download patches made by other composers.
    'Projects | Cycling '74' <https://cycling74.com/projects> [accessed 15 March 2024].
    IRCAM, 'Home | Ircam Forum' <https://forum.ircam.fr/> [accessed 15 March 2024].
32 Artichoke, 'Lumiere | Light Art Festival | November 2023', *Lumiere Festival* <https://www.lumiere-festival.com/> [accessed 15 March 2024].
33 An estimated 160,000 people attended across the Lumiere festival in 2023. Brera-London and Artichoke, 'Lumiere 2023 Wrap Release' <https://www.lumiere-festival.com/media-hub/> [accessed 25 March 2024].
34 Daily tous les jours, *21 Balançoires (21 Swings) | Daily tous les jours* <https://www.dailytouslesjours.com/en/work/21-swings> [accessed 15 March 2024].
35 Daily tous les jours, 'About | Daily tous les jours' <https://www.dailytouslesjours.com/en/about> [accessed 25 March 2024].
36 Sarah Atkinson and Helen W. Kennedy, 'From Conflict to Revolution: The Secret Aesthetic, Narrative Spatialisation and Audience Experience in Immersive Cinema Design', *Participations*, 13.1 (2016), 252–79.
37 It should be noted that due to the nature of Secret Cinema, participants were not able to film or record their experience.
38 This is already happening in archiving practices for video games and can be seen in museums collections such as the National Video Game Museum, 'Our Collecting' <https://thenvm.org/about/collecting/> [accessed 17 March 2024].
39 Google Arts & Culture, 'Street View: Tour Famous Sites and Landmarks' <https://artsandculture.google.com/project/street-view> [accessed 25 March 2024].
40 Open back headphones and speakers would blend with the audio characteristics of the space they're being played into.
41 John Cage, *Silence: Lectures and Writing by John Cage* (Hanover, NH: Wesleyan University Press, 1961), p. 128.

References

Atkinson, Sarah, and Helen W. Kennedy, 'From Conflict to Revolution: The Secret Aesthetic, Narrative Spatialisation and Audience Experience in Immersive Cinema Design', *Participations*, 13. 1 (2016), 252–279.
Brera-London and Artichoke, 'Lumiere 2023 Wrap Release' <https://www.lumiere-festival.com/media-hub/> [accessed 25 March 2024].
Busoni, Ferruccio, *Sketch of a New Esthetic of Music* (New York: Dover Publications Inc., 1962).
Cage, John, *Silence: Lectures and Writing by John Cage* (Hanover, NH: Wesleyan University Press, 1961).

Daily tous les jours, 'About | Daily tous les jours' <https://www.dailytouslesjours.com/en/about> [accessed 25 March 2024].

Goehr, Lydia, *The Imaginary Museum of Musical Works: An Essay in the Philosophy of Music* (Oxford: Clarendon Press, 1992).

Google Arts & Culture, 'Street View: Tour Famous Sites and Landmarks' <https://artsandculture.google.com/project/street-view> [accessed 25 March 2024].

Hamilton, Andy, and Lara Pearson, eds., *The Aesthetics of Imperfection in Music and the Arts: Spontaneity, Flaws and the Unfinished* (London, England: Bloomsbury Academic, 2020).

'Hatnote/Listen-to-Wikipedia' (hatnote, 2024) <https://github.com/hatnote/listen-to-wikipedia> [accessed 18 March 2024].

IRCAM, 'Home | Ircam Forum' <https://forum.ircam.fr/> [accessed 15 March 2024].

Linehan, Andy, 'British Library: Saving Our Sounds | British Council Music' <https://music.britishcouncil.org/news-and-features/2023-01-30/british-library-saving-our-sounds> [accessed 25 March 2024].

London Video Game Orchestra, 'London Video Game Orchestra' <https://www.lvgo.co.uk/our-mission> [accessed 15 March 2024].

McLaughlin, Moira E., 'Video Game Music as Art?', *Washington Post*, 21 July2011 <https://www.washingtonpost.com/lifestyle/style/video-game-music-as-art/2011/02/07/ABYW1SI_story.html> [accessed 15 March 2024].

Nowakowski, Teresa, 'Super Mario Bros., Madonna and More Join the National Recording Registry', *Smithsonian Magazine*, 2023 <https://www.smithsonianmag.com/smart-news/national-recording-registry-mario-bros-madonna-180981988/> [accessed 15 March 2024].

O'Hara, Meghan, 'Experience Economies: Immersion, Disposability, and Punchdrunk Theatre', *Contemporary Theatre Review*, 27. 4 (2017), 481–496, doi:10.1080/10486801.2017.1365289.

'Our Collecting', National Videogame Museum <https://thenvm.org/about/collecting/> [accessed 17 March 2024].

'Projects | Cycling '74' <https://cycling74.com/projects> [accessed 15 March 2024].

Royal Albert Hall, 'Video Games in Concert' <https://www.royalalberthall.com/tickets/events/2023/video-games-in-concert/> [accessed 15 March 2024].

Summers, Tim, *Understanding Video Game Music*. Cambridge University Press, 2016.

Yee, Nick, 'Gaming Motivations Group Into 3 High-Level Clusters', *Quantic Foundry*, 2015 <https://quanticfoundry.com/2015/12/21/map-of-gaming-motivations/> [accessed 12 March 2024].

Media and art examples

Artichoke, *Lumiere | Light Art Festival | November 2023* <https://www.lumiere-festival.com/> [accessed 15 March 2024].

Cage, John, 4' 33", (Original Version): Other Variations (London: Edition Peters, 1952).

Cage, John, Music of Changes, 1952, piano.

Cage, John, Variations III, 1962, for any one or any number of people performing any actions.

Daily tous les jours, *21 Balançoires (21 Swings) | Daily tous les jours* <https://www.dailytouslesjours.com/en/work/21-swings> [accessed 15 March 2024].

Hatnote, *Hatnote Listen to Wikipedia* <http://listen.hatnote.com/#> [accessed 15 March 2024].

Let's Play *LIVE! #1* - Super Mario Bros. *w/FULL ORCHESTRA!* Ft. George S*alazar*, dir. by The 8-Bit Big Band, 2018 <https://www.youtube.com/watch?v=2cRPzzZV8u0> [accessed 15 March 2024].

Super Mario Bros. *on* Marimba *(with 4* Mallets) *by* Aaron Grooves, dir. by AaronGrooves, 2015 <https://www.youtube.com/watch?v=3v7QC6Bl_7E> [accessed 15 March 2024].

Super Mario Bros. Symphony Performance | Game Awards *2020*, dir. by GameSpot, 2020 <https://www.youtube.com/watch?v=lIS9alKXvz8> [accessed 15 March 2024].

Super Mario Bros. Theme - Otamatone Cover, dir. by TheRealSullyG, 2020 <https://www.youtube.com/watch?v=gjDJjI82xtc> [accessed 15 March 2024].

Super Mario Bros. Theme *on* 2 Credit Card Machines, dir. by Device Orchestra, 2019 <https://www.youtube.com/watch?v=2NCviYMdV1I> [accessed 15 March 2024].

Chapter 7

Where now for interactive sound and music?

This final chapter intends to look at potentials for the future of interactive sound and music. As a practice, interactive work is still relatively young, but techniques and approaches are beginning to become solidified in how we create and develop interactive sound and music, as well as audience expectations.

With an industry so reliant on changes in audience expectations and on new and developing technologies, these structures and approaches may still be open to change and are very much open to the pushes from the market. This has never been evidenced so clearly as by the speculation about interactive experiences that appeared during the pandemic where the industry predicted that live music events wouldn't recover, being replaced by digital experiences, which is contradicted by the evident return to live music and immersive and interactive events.[1]

This chapter will take a speculative look at potential futures for interactive sound and music, based on current trends and approaches that are being introduced across multiple sectors. These will be used to consider how skills in composition are changing and what this will mean for aesthetics of interactive sound and music.

Digital experiences

Digital experiences continue to be an avenue where interactive sound and music thrives, through games platforms or through other exploratory digital platforms and events. Market research company Newzoo predicted that the number of gamers worldwide would reach 3.31 billion in 2023 (4.3% year on year growth), with mobile games being partially responsible for this growth.[2] 60% of the market play on mobile platforms, with 33% playing on PC and 32% playing on consoles.[3] During the pandemic games became one of the most popular activities for people in lockdown, offering an opportunity to socialise with friends and family at a distance.[4] This growth in the global games market suggests that games have become fully embedded in mainstream cultures, rather than being considered a fringe interest. This means that audiences that are already partially games literate will have a greater understanding of the types of interaction and exploration required within a game, supporting their

DOI: 10.4324/9781003344148-8

engagement and interaction in real world interactive sound installations and interactive works. This increase in the game market will also lead to an increased need for composers and sound designers with the skills to work in an interactive first approach, rather than interactive composition being considered an additional specialised skill for composers.

Alongside the increased use of games across platforms, games techniques and approaches are being used across other sectors providing access to game-like digital experiences. Museum experiences and digital archives have become more interactive; for example, the V&A's *David Bowie Is* exhibition launched an AR experience for people who could not attend the exhibition, or for those who wanted to explore the exhibits in more detail.[5] Similarly Google Arts & Culture offer virtual tours of 4984 museums.[6] These virtual exhibitions and archives also extend to more figurative representations of experiences, such as the V&A's Glastonbury archive which shows a representation of the Glastonbury site across time and includes changing soundscapes of the festival.[7] These online, digital experiences suggest the spread of game techniques into other sectors. The representation of realistic soundscapes within archiving, as demonstrated with the Glastonbury archive, suggests the importance of representing the full atmospheric experience of an event or an exhibition which will include implementing interactive audio techniques and may lead to an increase in binaural recording on location to better represent the audience experience.

While game techniques have been effectively implemented into online museum experiences and archiving approaches, they have not fully made it into streaming platforms used for more traditional film and television experiences. Netflix has made some attempts to embed interactive films into their platforms using a 'choose your own adventure format' as seen in Black Mirror: Bandersnatch[8], however, this has not yet reached wide stream use as, although some options are interactive, there are places where the films force you to remake a decision if it doesn't match the writers intended narrative arc or limits users so to ensure that they follow the overall narrative arc. This is most clearly demonstrated in Bear Grylls' *You vs. Wild,*[9] where users expressed frustration that Bear Grylls wouldn't make a bad decision no matter how much the user pushed.[10] These attempts from Netflix represent a recognition of how important interactivity will be to future users on their platform. The user responses also suggest that as we move towards more interactivity, audiences will expect their media to provide a real-time meaningful interaction where their input fully influences the outcomes of the work. Broadcast companies will need to look further into games techniques, reducing the amount of creative control that they exercise over their media products to allow for a more satisfying audience experience. For composers, even working across film and television, this will require them to have more flexible approaches to their compositions using game audio skills.

As the population becomes more at home with digital experiences, we can expect game audio techniques to begin to be used more widely across different media experience. Audiences are likely to expect these to be meaningful interactive experiences allowing playful exploration across different media types. Media composers will need to have interactive techniques as part of their core skillsets, allowing them to embed interactive approaches across these new digital experiences. The increase in digital interactive experiences will also have an impact on file sizes used for interactive sound and music, especially when being loaded onto mobile experiences. For example, with some earlier digital mobile experiences such as the Natural Trust's *Soho Stories*,[11] an audio tour app for London, the user had to download multi gigabyte sound files, taking up a large amount of mobile storage. Moving forward, as technology becomes more accessible and mobile, less memory-intensive solutions may be developed to support more mobile experiences.

Physical experiences

In person and community opportunities look likely to continue to grow as an area of interest and development. In an article exploring the future of the Experience Economy, Pine predicts that while these digital experiences will provide hybrid opportunities for people to engage with events, this will not replace physical and community experiences.[12] Post-pandemic, people's appetites for experience events increased, this was attributed by Pine to a shift in consumerism from owning more objects to gaining more social and live experiences that they missed during lockdowns.[13] This has led to more interactive and immersive approaches being added to what would have been traditional gallery or concert experiences.

For example, the Van Gogh immersive experience offers traditionally displayed prints of Van Gogh's work, deconstructions and sets built from his pieces so you can 'step into' the art, immersive projections of the work, the opportunity to display your work digitally in a gallery and a VR immersive experience. This exhibition has been shown in 48 cities globally, demonstrating the success of this more interactive format.[14] Another example is Dopamine Land, which offers a 'multisensory experience' in various interactive spaces including interactive musical tiles, ball pits and infinity rooms. These growing event experiences suggest that there is a public appetite for more interactive work that blends techniques from videos games with in-person, communal experiences. Many of these experiences exist as an experience in themselves without necessarily including context of wider artistic practices that they extend from, for example Dopamine Land's use of infinity rooms builds on the practices of Yayoi Kusama.[15] As these experience events continue to develop with shareable social media imagery and approaches taken from experimental art, it will be interesting to consider how this field begins to blend the artistic context for the work with the audience's expectations within the experience economy. These two things are not necessarily acting in conflict with each other and can be developed through presenting layers of experience and artistic meaning, where the initial impact of the work can be peeled away to discuss the further artistic intentions.

Immersive theatre appears to be moving from more specialised bespoke events, such as those created by Punchdrunk, to mainstream productions built on existing IP. These range from party-like productions with *The Great Gatsby* immersive production[16] to dining experiences like the *Mamma Mia Party*[17] providing an updated version of dinner theatre. Some productions are creating a hybrid of immersive and traditional theatre practices, as can be seen at The Bridge Theatre's production of *Guys and Dolls* which contains a traditional performance combined with a pit for audiences with an immersive ticket.[18] This merging of immersive experiences with traditional performance practices in theatre will have an impact on the skills required by sound designers working in theatre environments where audio requires a hybrid approach from traditional onstage audio to more subtle audio experiences embedded within the audience space.

Music festivals have also developed to include more interactive experiences, with Glastonbury 2023 featuring a range of interactive sound works including Silver Hayes' Pavilion, built to showcase sustainable fungi as a building material which housed an interactive sound work by Brian d'Souza, Or:La and Roisin Berkley.[19] This suggests that interactive sound installations could, in themselves, act as liminal spaces for larger concert experiences, especially in large-scale music festivals when there are periods between performances or times when people may be waiting for their preferred act or looking for a variation in experience; interactive spaces can provide an alternative within the schedule linked to the larger scale events.

The *ABBA Voyage* experience shows a popular use of interactive sound and music, in a performance context, through a blend of live sound and pre-recorded vocals from the band.[20] Within stadium concerts interactive LED wristbands, as used by Coldplay during their 2012 tour,[21] seem to have gained more widespread use across a range of concert experiences including Taylor Swift's Eras tour.[22] This continued trend in interactive and adaptive lighting design points towards artists looking to support musicking behaviours within their gigs, as audiences feel like they are not just watching the show but contributing to the overall experience and atmospheres of the event. These wristbands and similar technologies also provide shareable moments for social media, expanding the concepts of musicking into the virtual realm. These examples of interactive technology use across traditionally performance-led events point towards a potential change in the performer–audience hierarchy, with audiences taking a more participatory or collaborative role in the work and development of the atmosphere. As we move to future developments in interactive sound and music, this will lead to further use of interactive technology, furthering the audience's role in the performance.

The growing trend of experience events points to an audience appetite for interactive experiences. Combined with the increase in gamers, these future audiences will be aware of and seeking interactive experiences. This is likely to result in an increase of exploratory behaviours and a reduction in the formality of gallery and performance contexts. This suggestion is supported by Audience

Answers' market research showing that people are more likely to attend an event with relaxed behaviour rules, including taking photos and talking to others.[23] Interactive events moving forward will need to make a decision about the use of phones during experience events and the role that this plays in the immersion and engagement with the event versus the potentially increased attendance at more relaxed environments.[24]

Interactive sound and music in everyday life

As considerations change for everyday technologies, including looking for greener travel options, we are beginning to see more complex use of interactive sound and music within everyday technology design. For example, electric cars, that are now becoming more commonplace, are virtually silent as there is no internal combustion engine. This lack of sound presents a risk to pedestrians, particularly blind and partially sighted people.[25] In response to this, laws have been enacted to add artificial sounds, known as Audible Vehicle Alert Systems (AVAS), to cars to ensure pedestrian safety.[26]

In response to these changing requirements, car companies have been working with sound designers to implement sounds that are semiotically recognisable as vehicles, while reflecting the image of the company and reacting in real time to changes in the car activity. In particular, Nissan has considered their car engine sound to be a selling point of the vehicle. The sound is designed with cityscapes in mind and is localised for different countries, similar to game sounds processes for localisation.

Their engine sound works in four interactive phases with a start sound, an acceleration sound that reacts to the speed of the car, a continuous looping sound and a deceleration sound. This process is identical to the looping, adapting sound processes used in video games.[27] There is also a playfulness to some vehicle sound designs, with Tesla vehicles allowing users to upload their own bespoke sounds for functions such as the car horn.[28]

Similarly, these interactive audio techniques and aspects of audio as an interface are moving into considerations for healthcare and hospitals. Hospitals need to have a variety of alarms, which are constantly running to provide an audio signal to medical professionals about the status of a patient. While this is a necessary use of audio as an interface, the alarm and regular beeping sounds that have been used for years are considered to have a detrimental impact on audience wellbeing within a hospital space. Zhou et al. have been using VR systems to model the impact of sound on health and wellbeing in hospitals[29] while the composer Yoko Sen has been developing sound maps of the acoustic information in hospitals to build a better soundscape.[30] This builds from compositional traditions such as Eno's ambient music, which has been applied to hospital environments.[31]

These two examples of interactive systems in everyday life point to a growing trend of interactive sound and music moving out of arts and entertainment roles into more practical everyday use. While everyday technology does already contain elements of audio as an interface and is designed to show that the technology is working with citizen accessibility in mind, e.g. pedestrian crossing sounds or kitchen appliance sounds, these newer implementations of interactive sound and music point to more ambient-focused solutions that consider not just the practicality and effectiveness of the sound choices but also how pleasant they are for everyday life.

Separation between game and experimental music

As discussed in the introduction, the histories and approaches for interactive games, public art and experimental sound and music are usually discussed separately due to the differences in the practical functions of the work and the intended audiences. However, as audiences have access to different media and events across the experience economy, this separation appears to be shifting, allowing for greater sharing of approaches between fields.

The implementation of Max for Live in Ableton in 2009 was a significant shift in how different communities would start to overlap as it brought approaches and tools from Cycling '74s experimental environment into a more commercial live performance setting. Similar uses of experimental technology have been seen in other aspects of commercial and popular music, for example in Ariana Grande's use of Imogen Heap's MiMu gloves.[32] This suggests an appetite for more experimental music approaches in different fields.

Similarly, soundscapes in independent games seem to be shifting to experimental techniques. For example, the rhythm game *Thumper* uses techniques from minimalism and polyrhythms to create a complex and shifting rhythmic piece that the player needs to sync with in order to drive without collision.[33] Generative composition, similar to aleatoric techniques used by Cage in *Music of Changes*,[34] and generative techniques used by Eno in *Ambient 1: Music for Airports*[35] can be seen in games such as *Ape Out*, with its generative jazz soundtrack.[36]

More complex generative processes can be seen in *No Man's Sky*, where the planets and creatures are randomly generated.[37] Paul Weir's audio system creates a generative score and entirely generative processes are used to model creature sound and soundscapes. This suggests that there is a future space in more interactive games for the greater embedding of generative sound processes that provide the flexibility the game needs while reducing the amount of storage space for the audio.

The role of technology in interactive sound and music

Potential trends in technology will have a possible impact on interactive sound and music moving forward, including changing the hardware that people use when listening to audio and how they engage on a day-to day-basis with sound and music.

Binaural audio is looking likely to become more mainstream in audio practices, particularly as it can be implemented easily without audiences having to invest in specialised technology or headphones. Headphone usage is rising globally with increased production and consumption of headphones showing more personalised use of audio technology.[38] Haas et al. refer to this increased headphone use within the context of 'personal soundscape curation', where listeners use headphones to reduce atmospheric sounds and are also provided with the opportunity to curate their day through choices of music etc. Within the categories of sound identified by Haas et al. for headphone use was nature soundscapes for relaxation,[39] while Pine highlights that a key trend in the experience economy relates to health and wellbeing.[40] This points to what appears to be a rising trend in binaural beats, which benefit from being listened to through headphones and have some potential health and relaxation benefits as some preliminary research shows that they can have a positive effect on reducing stress anxiety.[41]

Additionally, head tracking technology consistent with the requirements of VR is becoming more widespread within every day technologies particularly as Apple have invested heavily in adding spatial audio across their headphone ranges, allowing for Dolby Atmos mixed tracks to be heard with additional spatial information for any users with the new Apple headphone products.[42] To ensure the effectiveness of the Head Related Transfer Function (HRTF), a number of companies, most notably Dolby, are investing in personalisation of HRTF calculations, taking into account the differences in people's ears and heads.[43] If spatial audio does get more widespread use, it will require techniques directly taken from game audio to develop effective audio mixes. In a game development platform, all sound is mixed from the perspective of a listener object, usually placed on the camera to represent the player in the space. In a spatial audio setting the listener object will be placed where the listener's head is, central to the headphones, with all sound being mixed in relation to this central listener object.

While VR technology has been reported to have steady growth,[44] it has not gained a widespread foothold in the home games market. Instead, VR can be seen more commonly within gallery experiences, professional training programmes (for example in the metaverse) and in experience event companies, such as VR games bars. Gallery and training uses will require interactive soundscapes that are extremely detailed and true to life while being interactive, requiring sounds design skills similar to those used in AAA games such as *Call of Duty*.

While the companies like the metaverse aim to offer social experiences, VR for home use is not adaptable to sofa co-op or social play in person, which limits the scope for more social or family-based users who aim to have an in-person experience alongside the game use. This is being recognised by companies as they pivot to AR and XR technology, most notably Apple's Vision Pro which aims to give a blended experience where users can experience the immersive technology offered by VR technology while being able to still engage with the space around them. Like Apple's headphones, the Vision Pro is compatible with spatial audio.[45]

While there has been much media coverage on the role of generative AI in a music context, this has often looked at the role of AI as a replacement for the composer and deals with the question of how data scraping has been used to collect musical examples to train AI systems, with artists expressing concern about the industry potentially replacing them with AI.[46] While this is a potential concern, if we look at these technological developments through the lens set by the experience economy, people are still looking for human connections within the art and media that they consume and they are still seeking transformative, personal experiences through their choice of events and media.[47] The examples of interactive sound and music discussed throughout this book have a shared element of human experience and empathy; even when the work is being demonstrated by a robot, in the case of Yuan and Yu.[48] Each of the pieces discussed throughout this book has the opportunity for an audience to discover more about the world or other people's experience. For example, while data generated, *Listen to Wikipedia* provides a fascinating insight into human behaviour through the edits they choose to make on Wikipedia every day.[49] This suggests to me that AI will not be used to replace composers, but it can present an interesting opportunity for the composer of interactive sound and music. If we use the definition of interaction as outlined in chapter 1, the instigator for interaction within a piece of work does not need to be human. Therefore, AI could become the instigator for an interactive piece of work, making it a collaboration with the composer based on conditions set by the composer. Practically speaking, AI could also support with the testing processes that a composer uses to ensure that a work is genetically complete and giving the composer more of an insight to the aesthetic completeness of a piece. So, while AI will not replace the composer, it could become a useful tool and provide the basis for interesting artistic experimentation.

Changing skills for composers and sound designers

All the potential avenues identified within this chapter suggest that the role of the composer and sound designer is changing what were traditional composition practices. While interactive techniques were previously seen as a specialism within the realms of games, public sound installations and experimental sound and music are now extending out to fields of environmental design and commercial music. Therefore, these interactive techniques need to be further embedded in the composer and sound designer's skillset to allow for the maximum flexibility and to understand audience requirements.

Composers across all areas need to be able to work using the flexible techniques outlined in chapter 3. They need to be able to develop musical ideas that can adapt and change with the work, and work using non-linear formats to adapt their work to the more interactive nature of media moving forward. In order to build realistic audio landscapes for VR training functions, they also need to build keen listening techniques where they can accurately represent real world soundscapes in as much detail as possible to represent not just the audio function in the real-world environment and the realism of the sound, but its place within the 3D space.

Audio engineers working on mixing and mastering will need to further develop their practice to fit with developing approaches for spatialisation where panning doesn't just work linearly but through the entire 3D sound space. This will also require similar skills to games about the user experience and locational sound.

Audio designers outside of traditional fictional interactive spaces will need to build and develop understandings of audio as an interface, building soundscapes that inform users of technology and instructions for use in an increasingly electronic world. While these have previously been very functional beeping sounds, noise pollution concerns and user experience suggests that composers need to build an understanding of the impact of their sounds on the population and the semiotic associations that can support the user experiences.

The changes in the role of the composer will have an impact on how composers are trained through higher education routes. While there are a number of specialised games audio courses being seen, particularly at a postgraduate specialism level, these interactive sound and music techniques are being further embedded into other compositional courses. The skills required in contemporary education for composition will be more varied and cross disciplinary, representing the varied skills required by the composer, as discussed in chapter 3.

The skills required in contemporary composition education will need to include:

- Traditional composition structuring, melodic development, harmonic techniques and orchestration
- Tools for structuring adaptive composition including building work in a non-linear format
- Understanding of audience behaviours and psychology, including how audiences listen within a space
- Understanding of sound within physical spaces
- Rudimentary coding skills including audio implementation in game engines and middleware[50]

Through this increasingly cross-disciplinary approach, the composer will be able to consider the holistic experience of the audience in line with the compositional needs of the work.

Conclusion

To conclude, while speculative, there are some key themes that are beginning to show for the future of interactive sound and music.

Experience events look like they will have a big impact on future exhibition approaches as users continue to have access to high quality home streaming. These events will require a range of interactive techniques, further embedding tools and approaches from games, public art and experimental sound and music. Audiences are likely to have greater expectations of these interactive experience events due to their increased understanding of and exposure to video game practices.

The role of audio as an interface will have a greater impact on our day-to-day life, with sound being designed and taking influences directly from video games to improve soundscapes in cities.

Experimental techniques are likely to be seen across other platforms through compositional and technological approaches, first in games but they have the potential to be seen in public art and within experience events.

Digital experiences will continue to expand with the metaverse taking on practical uses for training, requiring composers to build realistic 3D soundscapes using game audio techniques.

New technology will embed game audio techniques within personal audio experiences with binaural and spatial audio being built into personal listening habits, allowing for audiences to curate their personal experience within a public environment.

All of these trends have practical and artistic approaches that hold the audience at their centre. Traditionally the need of the musical work or the media is given priority when developing music. The musical process must serve the artistic needs of the piece or the narrative function of the work. Within the developments that we're seeing in interactive sound and music this balance is shifting to place the audience at the centre of the work. While we have traditionally worked with composer-based hierarchies it looks like this shift will move closer to the ideal of the audience as a collaborator in interactive sound and music.

Reading group questions

1   How do we model if predictions for future developments based on current trends are valid?

2   How could digital archives and museum experiences more accurately represent the real-world experience?

3   What would be the audio implications for a more flexible media experience using a format similar to Netflix's 'choose your own adventure' approach?

4   How can experience events be developed to provide layers of additional artistic meaning to the interactive work?

5 How could the performer/audience hierarchy shift in performance events by using techniques from interactive sound and music?

6 How could phone usage be fully embraced in an interactive sound and music event without impacting the overall immersive experience of the work?

7 If we were building a real-world audio interface from scratch, what decisions would need to be made when designing sounds for appliances and vehicles?

8 Which fields would potentially benefit from generative sound approaches and how could these be effectively incorporated?

9 How will changes in technology impact the sound of the music being created moving forward?

10 What approaches will composers need to use to ensure that they are able to adapt and futureproof their techniques for changes in interactive sound and music?

Notes

1 'Majority of Music Fans Embracing Full-Capacity Gigs and Taking Safety Precautions, Survey Shows', *Sky News* <https://news.sky.com/story/majority-of-music-fans-embracing-full-capacity-gigs-and-taking-safety-precautions-survey-shows-12384818> [accessed 11 April 2024].

2 Newzoo, *Global Games Market Report*, January 2024 <https://newzoo.com/resources/blog/games-market-estimates-and-forecasts-2023#:~:text=As%20per%20the%20January%202024,%2D5.1%25%20dip%20in%202022> [accessed 11 April 2024].

3 Newzoo, *How Consumers Engage with Video Games Today: Newzoo's Global Gamer Study 2023*, 2023 <https://newzoo.com/resources/trend-reports/global-gamer-study-free-report-2023> [accessed 8 April 2024].

4 Statista, 'Topic: COVID-19 Impact on the Gaming Industry Worldwide' <https://www.statista.com/topics/8016/covid-19-impact-on-the-gaming-industry-worldwide/> [accessed 11 April 2024].

5 The David Bowie Archive, Sony Music Entertainment (Japan) Inc., and Planeta, 'David Bowie Is – The AR Exhibition' <https://davidbowieisreal.com> [accessed 11 April 2024].

6 Google Arts & Culture, 'Street View: Tour Famous Sites and Landmarks' <https://artsandculture.google.com/project/street-view> [accessed 25 March 2024].

7 'V&A: Mapping Glastonbury', *V&A: Mapping Glastonbury* <https://www.vam.ac.uk/mapping-glastonbury/> [accessed 11 April 2024].

8 *Black Mirror: Bandersnatch*, dir. by David Slade (Netflix, 2018).

9 *You vs. Wild*, dir. by Ben Simms (Bear Grylls Ventures, Electus, Netflix, 2019).

10 Stuart Heritage, 'Damn You, Netflix! Why Won't You Let Me Kill Bear Grylls?', *The Guardian*, 10 April 2019 <https://www.theguardian.com/tv-and-radio/2019/apr/10/damn-you-netflix-why-wont-you-let-me-kill-bear-grylls-bandersnatch-extreme-survival> [accessed 11 April 2024].

11 Rethink Audio, *Soho Stories* (National Trust, 2012).

12 B. Joseph Pine, 'Exploring the Future of the "Experience Economy"', *Strategy & Leadership*, 51.1 (2023), 31–34 (p. 33) <https://doi.org/10.1108/SL-10-2022-0101>.

13 Pine, p. 32.

14 Fever Labs Inc., 'Van Gogh Exhibition: The Immersive Experience' <https://vangoghexpo.com/> [accessed 11 April 2024].

15 Yayoi Kusama, *Infinity Mirrored Room - Filled with the Brilliance of Life*, 2011.

16 *The Great Gatsby: The Immersive Show* <https://www.immersivegatsby.com/> [accessed 11 April 2024].

17 MM! The Party Limited, *Mamma Mia! The Party* <https://mammamiatheparty.co.uk/> [accessed 11 April 2024].

18 *Guys & Dolls*, Bridge Theatre <https://bridgetheatre.co.uk/whats-on/guys-and-dolls/> [accessed 11 April 2024].

19 Silver Hayes, 6° (Pavilion, Glastonbury, 2023) <https://www.glastonburyfestivals.co.uk/silver-hayes-introduces-a-new-artistic-pavilion-for-glastonbury-2023/> [accessed 11 April 2024].

20 ABBA, 'Hero Band - ABBA Voyage' <https://abbavoyage.com/the-concert/hero-band/> [accessed 11 April 2024].

21 Leonie Cooper, 'Coldplay Spent £4.22 Million on Colour Changing "Mylo Xyloto" Tour Wristbands', *NME*, 2013 <https://www.nme.com/news/music/coldplay-169-1238257> [accessed 11 April 2024].

22 Yunqi Li, 'The Tech behind Taylor Swift Concert Wristbands', *WIRED Middle East*, 2023 <https://wired.me/technology/the-tech-behind-taylor-swift-concert-wristbands/> [accessed 11 April 2024].

23 Audience Answers, 'Younger Audiences Prefer More Relaxed Behavioural Codes | Audience Answers' <https://evidence.audienceanswers.org/en/evidence/articles/younger-audiences-prefer-more-relaxed-behavioural-codes> [accessed 8 April 2024].

24 Although this enforcing of phone-free events seems to have supported engagement for companies such as Punchdrunk. Gemma Nettle, 'Punchdrunk Calls Time on Masked Work as It Announces Closure of The Burnt City', *The Stage*, 2023 <https://www.thestage.co.uk/news/punchdrunk-calls-time-on-masked-work-as-it-announces-closure-of-the-burnt-city> [accessed 11 April 2024].

25 RNIB, 'New Electric Vehicles Must Generate Sound', *RNIB* <https://www.rnib.org.uk/news/new-electric-vehicles-safety-milestone/> [accessed 10 April 2024].

26 UK Government, 'Commission Delegated Regulation (EU) 2017/1576 of 26 June 2017 Amending Regulation (EU) No 540/2014 of the European Parliament and of the Council as Regards the Acoustic Vehicle Alerting System Requirements for Vehicle EU-Type Approval (Text with EEA Relevance)', King's Printer of Acts of Parliament <https://www.legislation.gov.uk/eur/2017/1576/annex/paragraph/1> [accessed 10 April 2024].

27 Nissan Motor Corporation, 'Nissan Hits All the Right Notes with the New LEAF MY21 Featuring the "Canto" Sound', *Official Europe Newsroom*, 2021 <http://europe.nissannews.com/en-GB/releases/nissan-hits-all-the-right-notes-with-the-new-leaf-my21-featuring-the-canto-sound> [accessed 10 April 2024].

28 Nick Yekikian, 'Tesla Now Lets Owners Customize Their Cars' Horns', *MotorTrend*, 2021 <https://www.motortrend.com/news/tesla-boombox-mode-horn-sound/> [accessed 10 April 2024].

29 Zhou, Tianfu, Yue Wu, Qi Meng, and Jian Kang. 'Influence of the Acoustic Environment in Hospital Wards on Patient Physiological and Psychological Indices'. *Frontiers in Psychology* 11 (21 July 2020): 1600. <https://doi.org/10.3389/fpsyg.2020.01600>.

30 Yoko Sen, 'Sound Experience Research', *Sen Sound* <http://www.sensound.space/ser> [accessed 10 April 2024].

31 The Montefiore Hospital, 'The Montefiore Hospital Is the First Hospital in the World to Incorporate the Works of Brian Eno in the Design of the Building' <https://themontefiorehospital.co.uk/our-news/2013/05/22/the-montefiore-hospital-is-the-first-hospital-in-the-world-to-incorporate-the-works-of-brian-eno-in-the-design-of-the-building> [accessed 10 April 2024].

32 *Ariana Grande - Mimu Gloves and 'Why Try' (Live in Anaheim 4–10–15)*, dir. by TheRealConcertKing, 2015 <https://www.youtube.com/watch?v=1Kv2ozAJTOE> [accessed 11 April 2024].

33 *Thumper* (Drool, 2016).
34 John Cage, *Music of Changes*, 1952.
35 Brian Eno, *Ambient 1: Music for Airports* (E.G. Polydor, PVC, 1978).
36 Gabe Cuzzillo, *Ape Out* (Devolver Digital, 2019).
37 Hello Games, *No Man's Sky*.
38 'Headphones - Worldwide | Statista Market Forecast', *Statista* <https://www.statista.com/outlook/cmo/consumer-electronics/tv-radio-multimedia/headphones/worldwide> [accessed 11 April 2024].
39 Gabriel Haas, Evgeny Stemasov, and Enrico Rukzio, 'Can't You Hear Me?: Investigating Personal Soundscape Curation', presented at the 'MUM 2018: 17th International Conference on Mobile and Ubiquitous Multimedia, Cairo Egypt: ACM' 2018, pp. 59–69 (p. 65) <https://doi.org/10.1145/3282894.3282897>.
40 Pine, p. 32.
41 Katherine Kelton and others, 'The Efficacy of Binaural Beats as a Stress-Buffering Technique', *Alternative Therapies in Health and Medicine*, 27.4 (2021), 28–33.
42 Dylan Smith, 'Apple Music Unveils "Higher Royalty Value" for Spatial Audio Tracks', *Digital Music News*, 2023 <https://www.digitalmusicnews.com/2023/10/20/apple-music-spatial-audio-higher-royalty-value/> [accessed 11 April 2024].
43 'Personalized Head Related Transfer Function (PHRTF) - Dolby Professional' <https://professional.dolby.com/phrtf/> [accessed 11 April 2024].
44 Newzoo, *Global Games Market Report*.
45 Apple, 'Apple Vision Pro' <https://www.apple.com/apple-vision-pro/> [accessed 11 April 2024].
46 Artist Rights Alliance, '200+ Artists Urge Tech Platforms: Stop Devaluing Music', *Medium*, 2024 <https://artistrightsnow.medium.com/200-artists-urge-tech-platforms-stop-devaluing-music-559fb109bbac> [accessed 11 April 2024].
47 Pine.
48 Sun Yuan and Peng Yu, *Can't Help Myself* (Guggenheim Museum, 2016).
49 Hatnote, *Hatnote Listen to Wikipedia* <http://listen.hatnote.com/#> [accessed 15 March 2024].
50 I outline further implications and techniques for teaching interactive sound and music in *Teaching Electronic Music*. Lucy Ann Harrison, 'Teaching Principles of Interactive Sound: A Practice-Based Approach', *Teaching Electronic Music*, 2021, 90–102 <https://doi.org/10.4324/9780367815349-7>.

References

ABBA, 'Hero Band - ABBA Voyage' <https://abbavoyage.com/the-concert/hero-band/> [accessed 11 April 2024].
Alliance, Artist Rights, '200+ Artists Urge Tech Platforms: Stop Devaluing Music', *Medium*, 2024 <https://artistrightsnow.medium.com/200-artists-urge-tech-platforms-stop-devaluing-music-559fb109bbac> [accessed 11 April 2024].
Apple, 'Apple Vision Pro' <https://www.apple.com/apple-vision-pro/> [accessed 11 April 2024].
Audience Answers, 'Younger Audiences Prefer More Relaxed Behavioural Codes | Audience Answers' <https://evidence.audienceanswers.org/en/evidence/articles/younger-audiences-prefer-more-relaxed-behavioural-codes> [accessed 8 April 2024].
Cooper, Leonie, 'Coldplay Spent £4.22 Million on Colour Changing "Mylo Xyloto" Tour Wristbands', *NME*, 2013 <https://www.nme.com/news/music/coldplay-169-1238257> [accessed 11 April 2024].

Google Arts & Culture, 'Street View: Tour Famous Sites and Landmarks' <https://artsa ndculture.google.com/project/street-view> [accessed 25 March 2024].

Haas, Gabriel, Evgeny Stemasov, and Enrico Rukzio, '*Can't You Hear Me?: Investigating Personal Soundscape Curation*', presented at the 'MUM 2018: 17th International Conference on Mobile and Ubiquitous Multimedia, Cairo Egypt: ACM', 2018, pp. 59–69, doi:10.1145/3282894.3282897.

Harrison, Lucy Ann, 'Teaching Principles of Interactive Sound: A Practice-Based Approach', *Teaching Electronic Music*, 2021, 90–102, doi:10.4324/9780367815349-7.

'Headphones - Worldwide | Statista Market Forecast', Statista <https://www.statista. com/outlook/cmo/consumer-electronics/tv-radio-multimedia/headphones/worldwide> [accessed 11 April 2024].

Heritage, Stuart, 'Damn You, Netflix! Why Won't You Let Me Kill Bear Grylls?', *The Guardian*, 10 April 2019 <https://www.theguardian.com/tv-and-radio/2019/apr/10/dam n-you-netflix-why-wont-you-let-me-kill-bear-grylls-bandersnatch-extreme-survival> [accessed 11 April 2024].

Kelton, Katherine, Terri L. Weaver, Lisa Willoughby, David Kaufman, and Anna Santowski, 'The Efficacy of Binaural Beats as a Stress-Buffering Technique', *Alternative Therapies in Health and Medicine*, 27. 4 (2021), 28–33.

Li, Yunqi, 'The Tech behind Taylor Swift Concert Wristbands', *WIRED Middle East*, 2023 <https://wired.me/technology/the-tech-behind-taylor-swift-concert-wristbands/> [accessed 11 April 2024].

'Majority of Music Fans Embracing Full-Capacity Gigs and Taking Safety Precautions, Survey Shows', *Sky News* <https://news.sky.com/story/majority-of-music-fans-embra cing-full-capacity-gigs-and-taking-safety-precautions-survey-shows-12384818> [accessed 11 April 2024].

Nettle, Gemma, 'Punchdrunk Calls Time on Masked Work as It Announces Closure of The Burnt City', *The Stage*, 2023 <https://www.thestage.co.uk/news/punchdrunk-calls-tim e-on-masked-work-as-it-announces-closure-of-the-burnt-city> [accessed 11 April 2024].

Newzoo, 'Global Games Market Report', January 2024 <https://newzoo.com/resources/ blog/games-market-estimates-and-forecasts-2023#:~:text=As%20per%20the%20Janua ry%202024,%2D5.1%25%20dip%20in%202022> [accessed 11 April 2024].

Newzoo, 'How Consumers Engage with Video Games Today: Newzoo's Global Gamer Study 2023', 2023 <https://newzoo.com/resources/trend-reports/global-gamer-study-free-report-2023> [accessed 8 April 2024].

Nissan Motor Corporation, 'Nissan Hits All the Right Notes with the New LEAF MY21 Featuring the "Canto" Sound', *Official Europe Newsroom*, 2021 <http://europe.nissa nnews.com/en-GB/releases/nissan-hits-all-the-right-notes-with-the-new-leaf-my21-fea turing-the-canto-sound> [accessed 10 April 2024].

'Personalized Head Related Transfer Function (PHRTF) - Dolby Professional' <https://p rofessional.dolby.com/phrtf/> [accessed 11 April 2024].

Pine, B. Joseph, 'Exploring the Future of the "Experience Economy"', *Strategy & Leadership*, 51. 1 (2023), 31–34, doi:10.1108/SL-10-2022-0101.

RNIB, 'New Electric Vehicles Must Generate Sound' <https://www.rnib.org.uk/news/ new-electric-vehicles-safety-milestone/> [accessed 10 April 2024].

Sen, Yoko, 'Sound Experience Research', *Sen Sound* <http://www.sensound.space/ser> [accessed 10 April 2024].

Smith, Dylan, 'Apple Music Unveils "Higher Royalty Value" for Spatial Audio Tracks', *Digital Music News*, 2023 <https://www.digitalmusicnews.com/2023/10/20/apple-m usic-spatial-audio-higher-royalty-value/> [accessed 11 April 2024].

Statista, 'Topic: COVID-19 Impact on the Gaming Industry Worldwide' <https://www. statista.com/topics/8016/covid-19-impact-on-the-gaming-industry-worldwide/> [acces sed 11 April 2024].

The Montefiore Hospital, 'The Montefiore Hospital Is the First Hospital in the World to Incorporate the Works of Brian Eno in the Design of the Building' <https://themonte fiorehospital.co.uk/our-news/2013/05/22/the-montefiore-hospital-is-the-first-hospita l-in-the-world-to-incorporate-the-works-of-brian-eno-in-the-design-of-the-building> [accessed 10 April 2024].

UK Government, 'Commission Delegated Regulation (EU) 2017/1576 of 26 June 2017 Amending Regulation (EU) No 540/2014 of the European Parliament and of the Council as Regards the Acoustic Vehicle Alerting System Requirements for Vehicle EU-Type Approval (Text with EEA Relevance)', King's Printer of Acts of Parliament <https:// www.legislation.gov.uk/eur/2017/1576/annex/paragraph/1> [accessed 10 April 2024].

'V&A: Mapping Glastonbury' <https://www.vam.ac.uk/mapping-glastonbury/> [acces sed 11 April 2024].

Yekikian, Nick, 'Tesla Now Lets Owners Customize Their Cars' Horns', *MotorTrend*, 2021 <https://www.motortrend.com/news/tesla-boombox-mode-horn-sound/> [acces sed 10 April 2024].

Zhou, Tianfu, Yue Wu, Qi Meng and Jian Kang, 'Influence of the Acoustic Environment in Hospital Wards on Patient Physiological and Psychological Indices', *Frontiers in Psychology*, 11 (2020), 1600, doi:10.3389/fpsyg.2020.01600.

Media and art examples

Ariana Grande - Mimu Gloves and 'Why Try' (Live in Anaheim 4–10–15), dir. by TheRealConcertKing, 2015 <https://www.youtube.com/watch?v=1Kv2ozAJTOE> [accessed 11 April 2024].

Black Mirror: Bandersnatch, dir. by David Slade (Netflix, 2018).

Cage, John, Music of Changes, 1952, piano.

Cuzzillo, Gabe, *Ape Out* (Devolver Digital, 2019).

Eno, Brian, Ambient 1: Music for Airports (E.G. Polydor, PVC, 1978).

Fever Labs Inc., 'Van Gogh Exhibition: The Immersive Experience' <https://vangoghexp o.com/> [accessed 11 April 2024].

Guys & Dolls, Bridge Theatre.

Hatnote, *Hatnote Listen to Wikipedia* <http://listen.hatnote.com/#> [accessed 15 March 2024].

Hello Games, No Man's Sky.

Kusama, Yayoi, Infinity Mirrored Room - Filled with the Brilliance of Life, 2011.

MM! The Party Limited, *Mamma Mia! The Party* <https://mammamiatheparty.co.uk/> [accessed 11 April 2024].

Rethink Audio, *Soho Stories* (National Trust, 2012).

Silver Hayes, 6° (Pavilion, Glastonbury, 2023) <https://www.glastonburyfestivals.co.uk/ silver-hayes-introduces-a-new-artistic-pavilion-for-glastonbury-2023/> [accessed 11 April 2024].

The David Bowie Archive, Sony Music Entertainment (Japan) Inc., and Planeta, 'David Bowie Is – The AR Exhibition' <https://davidbowieisreal.com> [accessed 11 April 2024].

The Great Gatsby: The Immersive Show <https://www.immersivegatsby.com/> [accessed 11 April 2024].

Thumper (Drool, 2016).

You *vs.* Wild, dir. by Ben Simms (Bear Grylls Ventures, Electus, Netflix, 2019).

Yuan, Sun, and Peng Yu, *Can't Help Myself* (Guggenheim Museum, 2016).

Index

Page numbers in **Bold** refers to tables and page numbers in *Italics* refer to figures

Printed in the United States
by Baker & Taylor Publisher Services